EDIB GARDENING *for* WASHINGTON *and* OREGON

Vegetables, Herbs, Fruits & Seeds

Marianne Binetti and Alison Beck

Lone Pine Publishing International

First printed in 2010 10 9 8 7 6 5 4 3 2 1
Printed in China

The Distributor: Lone Pine Publishing
1808 B Street, Suite 140
Auburn, WA USA 98001
Website: www.lonepinepublishing.com

Publisher's Cataloging-In-Publication Data
(Prepared by The Donohue Group, Inc.)

Binetti, Marianne, 1956-
Edible gardening for Washington and Oregon / Marianne Binetti and Alison Beck.

p. : ill., maps ; cm.

Includes index.
ISBN: 978-976-650-048-1

1. Gardening--Washington (State) 2. Gardening--Oregon. 3. Plants, Edible--Washington (State) 4. Plants, Edible--Oregon. 5. Plant selection--Washington (State) 6. Plant selection--Oregon. I. Beck, Alison, 1971- II. Title.

SB450.97 .B56 2010
635/.09797

Photos: All photos by Laura Peters and Nanette Samol, except: Joan de Grey 189b; Franky De Meyer 37a, 87a; Tamara Eder 10, 31c, 32b, 35bc, 36a, 39a, 42ab, 43,47b, 78, 91, 147a, 173ab, 175ab, 197a; Elliot Engley 26abc, 27ab; Jen Fafard 94, 117, 217a; Derek Fell 38ab, 56, 57ab, 101b, 102, 103, 107a, 131a, 135a, 150, 161a, 171a, 195; Erica Flatt 11a; Liz Klose 40a, 55b, 69a, 81, 86, 112, 168a, 177b; Trina Koscielnuk, 106; L. Lauzuma 49, 187; Scott Leigh 99; Janet Loughrey 190b; Heather Markham 47c; Tim Matheson 15a, 17a, 19abcd, 20, 21ab, 24bcd, 29d, 32a, 36b, 40b, 41ab, 45, 46, 47a, 48, 52, 53a, 90, 105b, 132, 133b, 214, 215b, 216ab, 217bc; Marilynn McAra 156a; Kim O'Leary 53b; photos.com 3, 28c, 35a, 55a, 60, 61a, 74, 75ab, 76, 83a, 97ab, 229, 233a; Robert Ritchie 39b, 133a; Sandy Weatherall 8b, 87b, 140, 151, 204; Don Williamson 17b.

PC: 15

Contents

Introduction

Edible gardens have been part of human culture and the Native American tradition in Washington and Oregon for thousands of years. Along with harnessing fire, developing the wheel and domesticating animals, cultivating food is one of the benchmarks of human advancement. Growing plants that provide food and learning to store that food for times of scarcity were advancements that allowed humans to develop civilizations.

Here in the Pacific Northwest, we have been leaders in protecting the environment, and growing food in urban parking strips and suburban backyards has become a practical solution to not only healthier eating but a healthier environment as well. When we buy food from the grocery store, we give little thought to where it came from, how far it had to travel and how much it cost to transport it, first to the store and then to the table. We don't think about how it was grown or who grew it. When we grow our own food plants, we develop a greater appreciation for the food, our gardens and our own ability to provide for ourselves and our families.

An edible garden includes much more than just vegetables. In Washington and Oregon we can grow a multitude of edible fruits, seeds, flowers and herbs in our gardens. You can start by adding a few favorites to patio containers and borders, or you can design an entire landscape with edibles—the possibilities are endless.

There are many reasons to grow edible plants. You can save money on your grocery bill; it's environmentally friendly because fewer resources are used for growing and getting the food to your table; it saves wildlife habitat by reducing the need to expand cultivated land; it

Rhubarb and hops grace a sunny corner.

gives you a chance to seek out varieties and specialty items; and an often-overlooked reason to grow edible plants is that they are attractive and often unique in appearance.

Many gardeners are put off by the thought of digging up a square or rectangle and having an uninspiring display of rows in their carefully landscaped garden. Edible gardening doesn't have to be that way. You can add plenty of edible plants to the landscape you already have. They don't have to be planted in rows; single plants and small groups can make attractive features. Also, if you do not plant a single type of plant in one location, pest and disease problems are less likely to affect your entire crop.

Think about your favorite vegetables and fruits. You don't have to try to accommodate your food needs for an entire year, but you could supplement what you buy by growing a few plants. Do you eat a lot of broccoli? Three to six plants can provide two people with a lot of meals because many selections produce additional smaller heads once the main one is cut. A container of tomatoes combined with flowering annuals on a sunny balcony or deck is beautiful and functional. Four zucchini plants will leave you wondering what to do with all your extra zucchini.

The climate and growing season in Washington and Oregon varies dramatically because of the influence of the Pacific Ocean along the coast, Puget Sound near Seattle and the dominating Cascade mountain range. In general, the climate is mild west of the Cascades,

especially near Puget Sound and Seattle and along the coast, where the marine influence keeps winters and summers mild. The growing season can last from 150 to 200 days. Winter temperatures rarely fall below 28° F, so harvesting from a winter garden of root crops and even lettuce and other greens is possible—especially with the protection of a cold frame or heavy mulch.

Eastern Washington and Oregon have a much more dramatic climate, with a shorter growing season but more intense summer heat. Fruit trees do especially well as you move toward the mountains, especially in the Willamette and Columbia River valleys. The growing season is as short as 115 days in the higher elevations on the eastern side of the mountains, but there is plenty of sunshine, so there is a wide choice of fruits and vegetables that do well—if you choose varieties carefully for cold tolerance.

Edible plants that do not do well in Washington and Oregon are those that need a long growing season with warm nights. Crops that typically mature quickly in the southern states, such as heat-loving okra and melons, are a challenge to grow in Oregon and Washington. Leafy crops, root crops, berries, beans and peas love our climate, and even heat-loving tomatoes, squash and corn do well when the right varieties are chosen. Ultimately, gardeners in Washington and Oregon will have a bushel basket of edible plants to choose from, and the more plants and varieties you experiment with growing, the better your chance of discovering just what grows best in the unique micro-climate and soil of your own little patch of Eden.

"If in doubt, test it out" should be the motto of gardeners all over Washington

Hardiness Zones

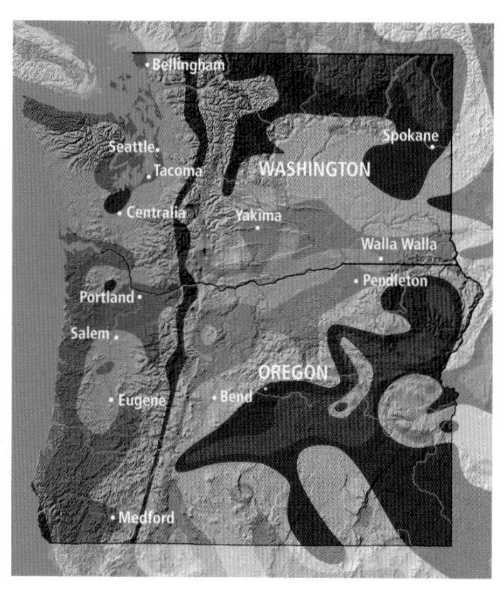

TEMPERATURE (°C)	ZONE	TEMPERATURE (°F)
–28.9 to –31.6	4b	–20 to –25
–26.2 to –28.8	5a	–15 to –20
–23.4 to –26.1	5b	–10 to –15
–20.6 to –23.3	6a	–5 to –10
–17.8 to –20.5	6b	0 to –5
–15.1 to 17.7	7a	5 to 0
–12.3 to 15.0	7b	10 to 5
–9.5 to –12.2	8a	15 to 10
–6.7 to–9.4	8b	20 to 15
–3.9 to –6.6	9a	25 to 20
–1.2 to –3.8	9b	30 to 25

and Oregon, as we are lucky to have local seed companies that specialize in fruits and vegetables for our unique climate. Ed Hume Seeds of the Seattle area has a long tradition of seed selection and growing tips for Washington and Oregon's cool growing season, and Raintree Nursery near the Cascade mountains specializes in fruits, nuts and berries and will ship plant starts right to your door. In Oregon, Nichols Seed Company caters to the gourmet cook and gardener, while Territorial Seeds has varieties and growing information that support year-round gardening west of the Cascades with organic and heirloom seeds and plant starts.

The three most important pieces of climate information for Washington and Oregon gardeners are the hardiness zone, the last frost date of spring and the first frost date of fall. These rules are not hard and fast, but they are excellent guidelines to help you pick plants and plan your garden. Hardiness zones (see map, above)

Frost Free Days

WASHINGTON/OREGON
LAST SPRING FROST DATES

- MARCH 15 - APRIL 1
- APRIL 1 - APRIL 15
- APRIL 15 - MAY 1
- MAY 1 - MAY 15
- MAY 15 - JUNE 1
- JUNE 1 - JUNE 15

WASHINGTON/OREGON
FIRST FALL FROST DATES

- AUGUST 1 - AUGUST 15
- AUGUST 15 - SEPTEMBER 1
- SEPTEMBER 1 - SEPTEMBER 15
- SEPTEMBER 15 - OCTOBER 1
- OCTOBER 1 - OCTOBER 15
- OCTOBER 15 - NOVEMBER 1

are relevant mostly for perennials, trees and shrubs. Plants are rated based on the hardiness zones they will grow in. The frost dates are a good estimate of season length (see map, above). If you can depend only on four frost-free months, you may want to choose plants that will survive a light frost or that will mature during your anticipated growing season.

This information provides a good starting point but should not completely rule your planting decisions. An early, warm spring is an excellent opportunity to set a few plants out, while a cold, wet, late spring might mean waiting a week or two later than usual.

As with any gardening, growing edible plants should be fun. They can add unique colors and textures to your garden and your dinner plate. Experiment with a few each year, and you may find yourself looking for space to add even more.

Edibles in the Garden

Edible gardens come in many forms and include a wide variety of plants. The neat rows of a traditional vegetable garden were adopted from the farm garden, and the attention paid to plant and row spacing is designed to make large numbers of plants more easily accessible. If you have plenty of space and want loads of vegetables to store over winter, this style can work for you.

There is no need to segregate edible plants from ornamental ones. The French potager, or kitchen garden, is a garden that is both decorative and functional. It generally consists of a symmetrical arrangement of raised beds. Plants are often repeated in a location in the bed rather than having one bed of all the same plant. Vegetables are combined with herbs and fruiting shrubs as well as edible flowers.

In intensive gardens, plantings are made in groups rather than rows. They may be formal or informal. An example of a more formal intensive garden is a square-foot garden. A square raised bed measuring 4' along each side is divided into 16 smaller planting squares. Each square is planted with as much of a single crop as the space will allow. As soon as a crop has matured and been harvested, something new is planted to replace it as long as the growing season allows.

An informal intensive garden could resemble a cottage-style garden, with vegetables planted in groups and drifts throughout the beds. Tucking groups of vegetables into existing beds could be formal or informal, depending on the nature of the garden you have.

There are many options for gardeners with limited gardening space. Many

fruits, vegetables and herbs grow successfully in containers on a balcony, porch, patio or deck. If you don't have room for trailing vines to spread, consider a trellis or obelisk and have the vines grow up instead of out. Many of the same vine-forming plants that grow up trellises will also trail nicely over the edge of a hanging basket.

As your garden grows and develops and you need to add new plants, think about adding edible ones. If you need a new shrub, perhaps you could include a raspberry or blueberry bush; asparagus, fiddlehead fern and rhubarb are hardy perennial choices; and the wealth of annual options is nearly limitless.

Ornamental Value

The ornamental value of edible plants is too often overlooked. Edible plants vary in size, color and texture and often have wonderful flowers, stunning foliage or decorative fruit in colors and shapes not often replicated in our ornamental plants. When you look at an edible plant, don't just think about the end result; think about the appearance of the plant and what it can add to your ornamental garden.

When planting intensively, keep plants fairly close together but leave enough room for each one to grow and spread. Many edibles grow from small seedlings into large, mature plants very quickly, and your yield will be greatly reduced if plants don't have at least some room to spread.

Lettuce (above), fiddlehead (center), red cabbage (below)

Many edible plants have stunning foliage. Their leaves come in a wide range of colors, from bloomy gray-green and greens so dark they're almost black to shades of red, purple, yellow, blue and bronze. Some foliage is patterned or has veins that contrast with the color of the leaves. Foliage can be immense or delicate and feathery.

Edibles with Interesting Foliage

Artichokes
Asparagus
Cabbage
Carrots
Chard
Coriander
Fennel (invasive in Washington)
Fiddlehead Ferns
Kale
Leeks
Lettuce
Mustard
Rhubarb
Squash

Getting Started

Finding the right edibles for your garden requires experimentation and creativity. Before you start planting, consider the growing conditions in your garden; these conditions will influence not only the types of plants you select, but also the location in which you plant them. Plants will be healthier and less susceptible to problems if grown in optimum conditions. It is difficult to significantly modify most of your garden's existing conditions; an easier approach is to match the plants to the garden.

Your plant selection will be influenced by the levels of light in your garden; the porosity, pH and texture of the soil; the amount of exposure; and the plants' frost tolerance. Sketching your garden may help you visualize how various conditions might affect your planting decisions. Note shaded areas, low-lying or wet areas, exposed or windy sections, etc. Understanding your garden's growing conditions will help you learn where plants will perform best and prevent you from making costly mistakes in your planning.

Light

There are four basic levels of light in a garden: full sun, partial shade (partial sun), light shade and full shade. Buildings, trees, fences and the position of the sun at different times of the day and year affect available light. Knowing what light is available in your garden will help you determine where to place each plant.

Plants in full sun locations, such as along south-facing walls, receive more than six hours of direct sunlight during the day. Locations classified as partial shade, such as east- or west-facing walls, receive direct sunlight for part of the day (four to six hours) and shade for the rest. Light shade locations receive shade for most or all of the day, though some sunlight does filter through to ground level. An example of a light shade location might be the ground under a small-leaved tree such as a birch. Full shade locations, which can include the north side of a house, receive no direct sunlight.

Plant your edibles where they will grow best. If your garden has hot, dry areas or low-lying, damp places, select plants that prefer those conditions. Experimenting will help you learn about the conditions of your garden.

Raised bed in full sun

Soil

Soil quality is an extremely important element of a healthy garden. Plant roots rely on the air, water and nutrients that are held within the soil. Of course, plants also depend on soil to hold them upright. The soil in turn benefits from plant roots breaking down large clumps while preventing erosion by binding together small particles and by reducing the amount of exposed surface. When plants die and break down, they add organic nutrients to soil and feed beneficial microorganisms.

Soil is made up of particles of different sizes. Sand particles are the largest—water drains quickly from sandy soil, and nutrients tend to get washed away. Sandy soil does not compact very easily because the large particles leave air pockets between them. Clay particles, which are the smallest, can be seen only through a microscope. Clay holds the most nutrients, but it also compacts easily and has little air space. Clay is slow to absorb water and equally slow to let it drain. Silt is midway between sand and clay in particle size. Most soils are a combination of these three particles and are called loams.

It is important to consider the pH level (the scale on which acidity or alkalinity is measured) of soil, which influences the availability of nutrients. Most plants thrive in soil with a pH between 5.5 and 7.5. Soil pH varies a great deal from place to place. West of the Cascades, the soil is naturally acidic, meaning that edible plants such as blueberries do well, but you will need to supplement almost all soils in western Washington and western Oregon with calcium and dolomite lime for peak fertility. You can buy testing kits at most garden centers, and soil-testing labs can analyze the pH and the quantities of various nutrients in your soil. For plants that prefer a pH that varies greatly

Sunflowers tolerate a range of soil conditions (above); chard (below) is one of the few vegetables that tolerates light shade.

from that of your garden soil, use planters or create raised beds where it is easier to control and alter the pH level of soil. The acidity of soil can be reduced with the addition of horticultural lime or wood ashes and increased with the addition of sulfur, peat moss or pine needles.

Water drainage is affected by soil type and the terrain in your garden. Plants that prefer well-drained soil and do not require a large amount of moisture grow well in a hillside garden with rocky soil. Water retention in these areas can be improved through the addition of organic matter. Plants that thrive on a consistent water supply or in boggy conditions are ideal for low-lying areas that retain water for longer periods or that hardly drain at all. In extremely wet areas, you can improve drainage by adding gravel, creating raised beds or using French drains or drainage tile.

Exposure

Your garden is exposed to wind, heat, cold and rain. Some plants are better adapted than others to withstand the potential damage of these forces. Buildings, walls, fences, hills, hedges, trees and even tall perennials influence and often reduce exposure.

Wind and heat are the elements most likely to cause damage. The sun can be very intense, and heat can rise quickly on a sunny afternoon. Choose edibles that tolerate or even thrive in hot weather for your garden's hot spots.

Too much rain can damage plants, as can over-watering. Early in the season, a light mulch will help prevent seeds or seedlings from being washed away in heavy rain. Most established plants beaten down by heavy rain will recover, but some are slower to do so. Waterlogged soil can encourage root rot

Open spaces expose plants to stronger wind.

because many edible plants prefer well-drained soil.

Hanging moss-lined baskets are susceptible to wind and heat exposure, losing water from the soil surface and the leaves. Hanging baskets look wonderful, but watch for wilting, and water the baskets regularly to keep them looking great. New water-holding polymers that hold water and release it as the soil dries have been developed for use in soil mixes.

Mulch (above)

Frost Tolerance

When choosing edibles, consider their ability to tolerate an unexpected frost. Most gardeners here can expect a chance of frost until early to mid-May, though warmer areas may have a last frost date in April and colder areas in June. The map on p. 9 gives a general idea of when you can expect your last frost date. Your local garden center should be able to provide more precise information on frost expectations for your area.

Edible plants are grouped into three categories based on how tolerant they are of cold weather: hardy, half-hardy or tender.

Hardy plants tolerate low temperatures and even frost. They can be planted in the garden early and may survive long into fall or even winter. These plants often fade in the heat of summer after producing an early crop. Many hardy edibles are sown directly in the garden in the weeks before the last frost date and can be sown again in summer for a second crop in fall.

Half-hardy plants can tolerate a light frost but will be killed by a heavy one. These edibles can be planted out around the last frost date and will generally benefit from being started indoors from seed if they are slow to mature.

Kale tolerates freezing temperatures (above). Lettuce tolerates light frost, while nasturtiums are quickly killed by cold (below).

Tender plants have no frost tolerance at all and might suffer if the temperature drops to even a few degrees above freezing. These plants are often started early indoors and are not planted in the garden until the last frost date has passed and the ground has had a chance to warm up. The advantage is that these edibles are often tolerant of hot summer temperatures.

Protecting plants from frost is relatively simple. Cover plants overnight with sheets, towels, burlap or even cardboard boxes—don't use plastic because it doesn't provide any insulation.

Peas prefer cold weather and will tolerate spring frosts.

Choosing the type of edibles to grow is just the beginning. There are often dozens of selections of some of our favorites. Choose plants and cultivars that are suitable to the climate and conditions in your garden. For example, avoid pumpkin, which needs four or five months to mature, if you can only be sure of three frost-free months.

Preparing the Garden

Taking the time to properly prepare your garden beds will save you time and effort throughout summer. Starting out with as few weeds as possible and with well-prepared soil that has had organic material added will give your edibles a good start. For container gardens, use potting soil because regular garden soil loses its structure when used in pots, quickly compacting into a solid mass that drains poorly.

Loosen the soil with a large garden fork and remove the weeds. Avoid working the soil when it is very wet or very dry because you will damage the soil structure by breaking down the pockets that hold air and water. Add good quality compost and work it into the soil with a spade, fork or rototiller. To determine how much compost you need, measure the area of your garden and then calculate how much compost you would need to cover it with a 2–4" layer.

Organic matter is a small but important component of soil. It increases the water-holding and nutrient-holding capacity of sandy soil and binds together the large particles. In a clay soil, organic matter will increase the water-absorbing and draining potential by opening up

spaces between the tiny particles. Common organic additives for your soil include grass clippings, shredded leaves, peat moss, chopped straw and well-rotted manure.

Composting

Any organic matter you add will be of greater benefit to your soil if it has been composted first. In natural environments, compost is created when leaves, plant bits and other debris are broken down on the soil surface. This process will also take place in your garden beds if you work fresh organic matter into the soil. However, microorganisms that break down organic matter use the same nutrients as your plants. The tougher the organic matter, the more nutrients in the soil will be used trying to break the matter down, thus robbing your plants of vital nutrients, particularly nitrogen. Also, fresh organic matter and garden debris might encourage or introduce pests and diseases to your garden.

A compost pile or bin, which can be built or bought, creates a controlled environment where organic matter can be fully broken down before being introduced to your garden. Good composting methods also reduce the possibility of spreading pests and diseases.

Creating compost is a simple process. Vegetable kitchen scraps, grass clippings and fall leaves will slowly break down if left in a pile. You can speed up the process by following a few simple guidelines:

- Put both dry and fresh materials into your compost, with a larger proportion of dry matter such as chopped straw, shredded leaves or sawdust. Fresh green matter, such as vegetable scraps, grass clippings or pulled weeds, breaks down quickly and produces nitrogen, which feeds the decomposer organisms while they break down the tougher dry matter.
- Layer the green matter with the dry matter, and mix in small amounts of soil from your garden or previously finished compost—it will introduce beneficial microorganisms. If the pile seems very dry, sprinkle on some water—the compost should be moist but not soaking wet, like a wrung-out sponge.

- Every week or two, turn the pile over or poke holes into it. Aerating the material will speed up decomposition. A compost pile that is kept aerated can generate a lot of heat. Temperatures can reach 160° F or more. Such high temperatures destroy weed seeds and kill many damaging organisms. Most beneficial organisms are not killed until the temperature rises higher than 160° F. To monitor the temperature of the compost near the middle of the pile, you can buy a thermometer attached to a long probe, similar to a large meat thermometer. Turning your compost when the temperature reaches 160° F will stimulate the process to heat up again, but will prevent the temperature from becoming high enough to kill the beneficial organisms.
- Don't put diseased or pest-ridden materials into your compost pile. If the damaging organisms are not destroyed, they could spread throughout your garden.
- When you can no longer recognize the matter that you put into the compost bin, and the temperature no longer rises upon turning, your compost is ready to be mixed into your garden beds. Getting to this point can take as little as one month and will leave you with organic material that is rich in nutrients and beneficial organisms.

Compost can also be purchased from most garden centers, and you can also practice the passive method of composting—just pile it up and let it rot. A pile of garden debris 4' by 4' will yield compost in about six months even if you never turn the pile.

Selecting Edible Plants

Many gardeners consider the trip to the local garden center to pick out plants an important rite of spring, and many garden centers offer a few basic vegetable plants. Other gardeners find it rewarding to start their own plants from seed. Both methods have benefits, and you might want to use a combination of the two. Purchasing plants provides you with plants that are well grown, which is useful if you don't have the room or the facilities to start seeds. Some seeds require conditions that are difficult to achieve in a house, or they have erratic germination rates, which makes starting them yourself impractical. However, starting from seed offers you a far greater selection of species and varieties because seed catalogs often list many more plants than are offered at garden centers. Starting from seed is discussed starting on p. 25.

When browsing through a seed catalog, you may find references to hybrid and heirloom seeds. Hybrids are generally newer selections of plants. They have been bred for specific traits such as flavor, size, disease resistance or improved storability. Often developed for market growers and food exporters, hybrids usually have traits that make them suitable for packing and transporting long distances without spoiling. Many have also become favorites among home gardeners since their introduction. Hybrids rarely come true to type from collected seed.

Heirloom refers to plant selections that have been in cultivation for generations. Many gardeners like the connection with history, knowing that their grandparents grew the same plant. Some of the most intriguing vegetable selections are heirlooms, and many advocates claim the vegetables to be among the tastiest and most pest and disease resistant. Seeds can be collected from the plants, and offspring will be true to type.

Purchased plants are grown in a variety of containers. Some are sold in individual pots, some in divided cell-packs and others in undivided trays. Each type has advantages and disadvantages.

Plants in individual pots are usually well established and generally have plenty of space for root growth. These plants have probably been seeded in flat trays and then transplanted into individual pots once they developed a few leaves. The cost of labor, pots and soil can make this option expensive.

Plants grown in cell-packs are often inexpensive and hold several plants, making them easy to transport. There is less damage to the roots of the plants when they are transplanted, but because each cell is quite small, it doesn't take long for a plant to become root-bound.

Plants grown in undivided trays have plenty of room for root growth and can be left in the trays longer than in other types of containers, but their roots tend to become entangled, making the plants difficult to separate.

Regardless of the type of container, check for roots emerging from the holes at the bottom of the cells, or gently remove the plant from the container to look at the roots. If there are too many roots, the plant is too mature for the container, especially if the roots are wrapped around the inside of the container in a thick web. Such plants are slow to establish once they are transplanted into the garden.

The plants should be compact and have good color. Healthy leaves look

Hybrid corn stays sweet when picked.

Lemon cucumber is a tasty heirloom.

Healthy plants (left), unhealthy plants (above)

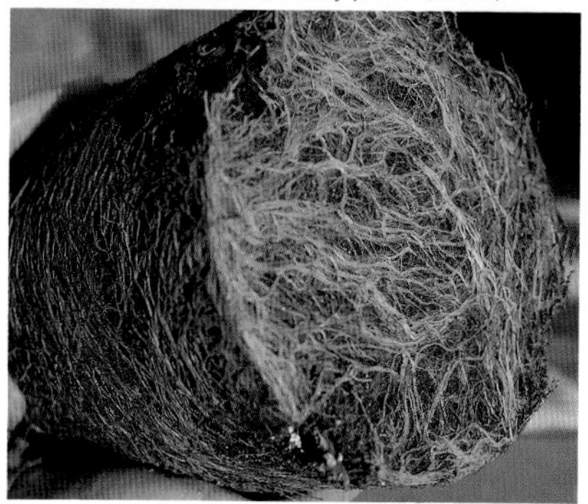

Root-bound plant (center), cutting roots before planting (below)

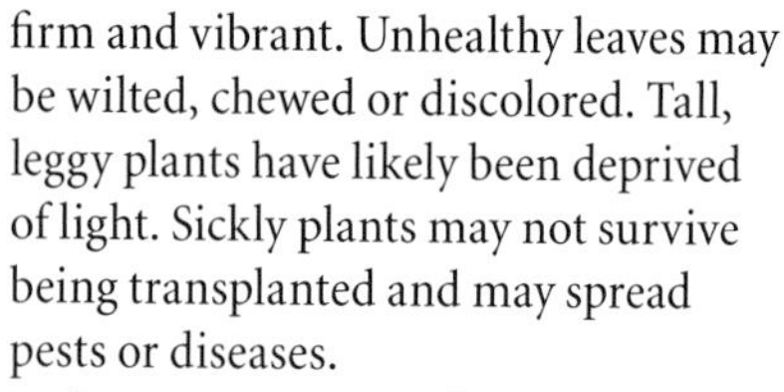

firm and vibrant. Unhealthy leaves may be wilted, chewed or discolored. Tall, leggy plants have likely been deprived of light. Sickly plants may not survive being transplanted and may spread pests or diseases.

Once you get your plants home, water them if they are dry. Plants growing in small containers may require water more than once a day. Begin to harden off the plants so they can be transplanted into the garden as soon as possible. Your plants are probably accustomed to growing in the sheltered environment of a greenhouse, and they will need to become accustomed to the outdoors. Placing them out in a lightly shaded spot each day and bringing them into a sheltered porch, garage or house each night for about a week will acclimatize them to your garden.

Nasturtium seeds (left), sunflower seeds (right)

Starting Edibles from Seed

Dozens of catalogs from different growers offer edible plants that you can start from seed. Several excellent seed companies in Washington and Oregon specialize in vegetable seeds especially for our climate. Ed Hume Seeds, Territorial Seed Company and Nichols Seeds are a few. Many gardeners spend their chilly winter evenings poring through seed catalogs and planning their spring and summer gardens. Other places to find seeds include the internet, local garden centers and seed exchange groups.

Starting your own plants can save you money, particularly if you need a lot of plants. The basic equipment is not expensive, and most seeds can be started in a sunny window. However, you may run out of room if you start with more than one or two trays. That is why many gardeners start a few specialty plants themselves but buy most plants from a garden center.

Each plant in this book has specific information on starting it from seed, if any is required, but a few basic steps can be followed for all seeds. The easiest way for you to start seeds is in cell-packs in trays with plastic dome covers. The cell-packs keep roots separated, and the tray and dome keep moisture in.

Seeds can also be started in pots, peat pots or peat pellets. The advantage to starting in peat pots or pellets is that you will not disturb the roots when you transplant them into the garden.

Why start early? Many edible plants need a long period of warm weather to mature. If the seed is sown directly into the garden after the risk of frost has passed, there might not be enough time for the plants to mature before the first fall frost ends the season.

Use a growing mix (soil mix) that is intended for seedlings. These mixes are very fine, usually made from peat moss, vermiculite and perlite. The mix will have good water-holding capacity and will have been sterilized to prevent pests and diseases from attacking your seedlings. One problem that can be caused by soil-borne fungi is damping off. The affected seedling will appear to have been pinched at soil level. The pinched area blackens, and the seedling topples over and dies. Using sterile soil mix, keeping soil evenly moist and maintaining good air circulation will prevent plants from damping off.

Fill your pots or seed trays with the soil mix and firm it down slightly. Soil that is too firmly packed will not drain well. Wet the soil before planting your seeds to prevent them from getting washed around. Large seeds can be planted one or two to a cell, but place smaller seeds in a folded piece of paper and sprinkle them evenly over the soil surface. Very tiny seeds can be mixed with fine sand and then sprinkled on the soil surface.

Small seeds do not need to be covered with any more soil, but medium-sized seeds can be lightly covered, and large ones can be poked into the soil. Some seeds need to be exposed to light to germinate; these should be left on the soil surface regardless of their size.

Additional light keeps plants healthy indoors.

Place pots or flats of seeds in plastic bags to retain humidity while the seeds are germinating. Many planting trays come with clear plastic covers that keep in the moisture. Remove the cover once the seeds have germinated.

Water seeds and small seedlings gently with a fine spray from a hand-held mister—small seeds can easily be washed around if the spray is too strong. At one greenhouse where the seed trays containing the annual flower sweet alyssum were once watered a little too vigorously, sweet alyssum was soon growing just about everywhere—with other plants, in the gravel on the floor and, oddly, in some of the flowerbeds outside. The lesson is, water gently.

Seeds provide all the energy and nutrients that young seedlings require. Small seedlings do not need to be fertilized until they have about four or five true leaves. When the first leaves that sprouted begin to shrivel, the plant has used up all its seed energy, and you can begin to use a fertilizer diluted to one-quarter.

If the seedlings get too big for their containers before you are ready to plant out, you may have to "up-pot" them to prevent them from becoming root-bound. Harden plants off by exposing

Potting up gives seedlings room to grow.

Broccoli (top), cucumber (center), watermelon (bottom)

them to outdoor conditions for longer every day for at least a week before planting them out.

Edibles to Start Indoors

Artichokes
Basil
Broccoli and other *Brassica* spp.
Cucumbers
Eggplant
Leeks
Melons
Peppers
Squash
Tomatoes

Some seeds can be planted directly in the garden. The procedure is similar to that of starting seeds indoors. Begin with a well-prepared bed that has been smoothly raked. The small furrows left by the rake help hold moisture and prevent the seeds from being washed away. Sprinkle the seeds onto the soil and cover them lightly with peat moss or more soil. Larger seeds can be planted slightly deeper. Very tiny seeds should be mixed with sand for more even sowing. Keep the soil moist to ensure even germination. Use a gentle spray to avoid washing the seeds around the bed because they inevitably pool into dense clumps. Cover your newly seeded bed with chicken wire, an old sheet or some thorny branches to discourage pets from digging in it.

Large seeds are easy to space out well when you sow them. With small seeds, you may find that the new plants need to be thinned out to give adjacent plants room to grow properly. Pull out the weaker plants when groups look crowded. Some are edible and can be used as spring greens in a salad or steamed as a side dish.

Edibles for Direct Seeding

Beans
Beets
Carrots
Chard
Coriander
Corn
Lettuce
Nasturtiums
Parsley
Peas
Radishes
Sunflowers

Beets

Lettuce

Radishes

Sunflower

Growing Edibles

Once your plants have hardened off, it is time to plant them out. If your beds are already prepared, you are ready to start. The only tool you are likely to need is a trowel. Be sure you have set aside enough time to do the job. You don't want to have young plants out of their pots and not finish planting them. If they are left out in the sun, they can quickly dry out and die. Try to choose an overcast day for planting.

Planting

Plants are easier to remove from their containers if the soil is moist. Push on the bottom of the cell to ease the plants out. If the plants were growing in an undivided tray, you will have to gently untangle the roots. If the roots are very tangled, immerse them in water and wash some of the soil away to free the plants from one another. If you must handle the plant, hold it by a leaf to avoid crushing the stems. Remove and discard any damaged leaves or growth.

The root ball should contain a network of white plant roots. If the root ball is densely matted and twisted, break the tangles apart with your thumbs to encourage the roots to extend and grow outward. New root growth will start from the breaks, allowing the plant to spread outward.

Plants started in peat pots and peat pellets can be planted pot and all. When planting peat pots into the garden, remove the top 1½–2½" of the pot. If any of the pot is sticking up out of the soil, it can wick moisture away from your plant.

Insert your trowel into the soil and pull it toward you, creating a wedge. Place your plant into the hole and firm the soil around it with your hands. Water gently but thoroughly. Until it is established, the plant will need regular watering.

Some of the plants in this book are sold in large containers. Plants in large containers can be planted as described

above, except that they may require larger holes. In a prepared bed, dig a hole that will accommodate the root ball. Fill the hole in gradually, settling the soil with water as you go.

Other plants may be sold as bare roots or crowns, or in moistened peat moss, sphagnum moss or sawdust. Plants that are sold without soil should be soaked in water for a few hours before planting as above, again being sure to accommodate the roots.

A few plants are sold as bulbs, such as garlic and onion sets, which can be planted about three times as deep as the bulb is high.

More detailed planting instructions are given, as needed, in the plant accounts.

Weeding

Controlling weed populations keeps the garden healthy and neat. Weeding can become a favorite task if you treat it as a soothing, reflective time in the garden—and good exercise if you gently stretch while you weed. Weeds compete with plants for light, nutrients and space, and they can also harbor pests and diseases.

Weeds can be pulled by hand or with a hoe. Shortly after a rainfall, when the soil is soft and damp, is the easiest time to pull them. A hoe scuffed quickly across the soil surface will uproot small weeds and sever larger ones from their

Mulch keeps weeds at bay.

roots. Try to pull weeds out while they are still small. Once they are large enough to flower, many will quickly set seed; then you will have an entire new generation to worry about.

Mulching

A layer of mulch around your plants prevents light from reaching weed seeds, keeping them from germinating. Those that do germinate will be smothered or will be unable to reach the surface, exhausting their energy before getting a chance to grow.

Mulch also helps maintain consistent soil temperatures and ensures that moisture is retained effectively. In areas that receive heavy wind or rainfall, mulch can protect soil and prevent erosion. Mulching is effective in garden beds and planters.

Organic mulches such as compost, grass clippings or shredded leaves add nutrients to the soil as they break down, thus improving the quality of the soil and, ultimately, the health of your plants.

Spread about 2–4" of mulch over the soil after you have finished planting, or spread your mulch first and then make spaces to plant afterward. Make sure the mulch is not piled thickly around the crowns and stems of your plants, especially in western Washington and Oregon where rain falls often. Mulch that is too close to plants traps moisture, prevents air circulation and encourages fungal disease. Replenish your mulch as it breaks down over summer. Use newspaper or cardboard beneath wood chips or hazelnut shells as a weed-blocking mulch for pathways in your garden. Placing flat boards in the rows can also serve as a mulch to walk on—and makes a great place to gather slugs and snails that attack your plants at night.

Watering

Water thoroughly but infrequently. Plants given a light sprinkle of water every day develop roots that stay close to the soil surface, making the plants vulnerable to heat and dry spells. Plants given a deep

watering once a week develop a deeper root system. In a dry spell, they will be adapted to seeking out water trapped deep in the ground. Use mulch to prevent water from evaporating out of the soil.

Be sure the water penetrates at least 4" into the soil; this is approximately 1" of applied water. To save time, money and water, you may wish to install an irrigation system, which applies the water exactly where it is needed, near the roots. Consult with your local garden centers or landscape professionals for more information.

Plants in hanging baskets and planters will probably need to be watered more frequently than plants in the ground—even twice daily during hot, sunny weather. The smaller the container, the more often the plants need watering.

A soaker hose provides a deeply penetrating application of water with less evaporation and run-off than a sprinkler.

Soaker hose arranged on a raised bed

Squash are heavy feeders.

Fertilizing

We demand a lot of growth and production from many of our edible plants; annual ones in particular are expected to grow to maturity and provide us with a good crop of fruit or vegetables. They, in return, demand a lot of sun, nutrients and water. Mixing plenty of compost into the soil is a good start, but fertilizing regularly can make a big difference when it comes time to harvest your crop.

Whenever possible, use organic or slow-release fertilizers because they are usually less concentrated and less likely to burn the roots of your plants. Organic fertilizers generally improve the soil as they feed your plants. Your local garden center should carry organic fertilizers. Follow the directions carefully—using too much fertilizer can kill your plants by burning their roots and may upset the microbial balance of your soil, allowing pathogens to move in or dominate.

Fertilizer comes in many forms. Liquids or water-soluble powders are easiest to use when watering. Slow-release pellets or granules are mixed into the garden or potting soil or are sprinkled around the plant and left to work over summer.

Organic amendments (left to right): moisture-holding granules, earthworm castings, glacial dust, mycorrhizae, bat guano, compost, bone meal and coir fiber.

Problems and Pests

Beneficial bugs: lacewing (top), ladybug (center), ladybug larvae (below)

New annual edible plants are planted each spring, and you may choose to plant different species each year. These factors make it difficult for pests and diseases to find their preferred host plants and establish a population. However, because many edibles are closely related, any problems that set in over summer are likely to attack all the plants in the same family.

For many years, pest control meant spraying or dusting with the goal to eliminate every pest in the landscape. A more moderate approach advocated today is IPM (Integrated Pest Management or Integrated Plant Management). The goal of IPM is to reduce pest problems so only negligible damage is done. Of course, you must determine what degree of damage is acceptable to you. Consider whether a pest's damage is localized or covers the entire plant. Will the damage kill the plant or is it affecting only the outward appearance?

Borage flowers attract beneficial insects.

Are there methods of controlling the pest without chemicals?

A good IPM program includes learning about the conditions your plants need for healthy growth, which pests might affect your plants, where and when to look for those pests and how to control them. Keep records of pest damage because your observations can reveal patterns useful in spotting recurring problems and in planning your maintenance regime.

There are four steps in effective and responsible pest management. Cultural controls are the most important and are the first response when problems arise. Physical controls should be attempted next, followed by biological controls. Resort to chemical controls only when the first three possibilities have been exhausted.

Cultural controls are the gardening techniques you use in the day-to-day care of your garden. Keeping your plants as healthy as possible is the best defense against pests. Growing plants in the conditions they prefer and keeping your soil healthy by adding plenty of organic matter are just two of the cultural controls you can use to keep pests manageable. Choose resistant varieties of plants that are prone to problems. Space the plants so that they have good air circulation around them and are not

stressed by competing for light, nutrients and space. Take plants that are destroyed by the same pests every year out of the landscape. Remove diseased foliage and branches, and burn them or take them to a permitted dumpsite. Prevent the spread of disease by keeping gardening tools clean and by tidying up fallen leaves and dead plant matter at the end of every growing season.

Physical controls are generally used to combat insect problems: picking insects off plants by hand, barriers that stop insects from getting to the plant, and traps that catch or confuse insects. Using floating row covers of lightweight fabric such as Reemay is the most effective barrier against most flying insects such as carrot rust fly and the white butterflies that lay eggs for cabbage worms. Physical control of disease often necessitates removing the infected plant part or parts to prevent the spread of the problem.

Biological controls use predators that prey on pests. Animals such as birds, snakes, frogs, spiders, lady beetles and certain bacteria can play an important role in keeping pest populations manageable. Encourage these creatures to take up permanent residence in your garden. A birdbath and birdfeeder will encourage birds to enjoy your yard and feed on a wide variety of insect pests. Many beneficial insects are probably already living in your landscape, and you can encourage them to stay by planting appropriate food sources. Many beneficial insects eat nectar from flowers such as coriander and perennial yarrow.

Chemical pesticides should rarely be necessary, but if you must use them, there are some organic options available. Organic pesticide sprays are no less dangerous than synthetic ones, but they break down into less harmful compounds.

Brussels sprouts (above), kohlrabi (below)

One of the most effective methods of reducing disease problems in an edible garden is crop rotation—don't put the same plant, or even one from the same family, in the same location two years in a row. By changing the crop from year to year, you can prevent problems from setting in annually. It is best to wait two years before planting any member of a family in the same location (the Brassica *family, for example, includes broccoli, Brussels sprouts, cabbage, cauliflower, collards, horseradish, kale, kohlrabi, mustard, oriental cabbage, radish, rutabaga and turnip).*

The main drawback to using any pesticide is that it may also kill the beneficial insects you have been trying to attract to your garden. Organic-based pesticides are available at most garden centers.

Follow the manufacturer's instructions carefully. A large amount of pesticide will not be any more effective in controlling pests than the recommended amount. Note that if a particular pest is not listed on the package, that product will not control that pest. Proper and early identification of pests is vital to finding a quick solution.

Whereas cultural, physical, biological and chemical controls are all possible defenses against insects, diseases can only be controlled culturally. It is most often weakened plants that succumb to diseases. Healthy plants can often fight off illness, though some diseases can infect plants regardless of their level of health. Prevention is often the only hope; once a plant has been infected, it should probably be destroyed to prevent the disease from spreading.

Insect damage disfigures asparagus spears (left); healthy spears (right) are more appetizing.

Aphids

Beetles

Alphabetical Guide to Pests and Diseases

Anthracnose

Fungus; yellow or brown spots on leaves; sunken lesions and blisters on stems; can kill plant.

What to do: choose resistant varieties and cultivars; keep soil well drained; thin out stems to improve air circulation; avoid handling wet foliage; remove and destroy infected plant parts; clean up and destroy debris from infected plants at end of growing season.

Aphids

Tiny, pear-shaped, green, black, brown, red or gray insects; can be winged or wingless, e.g., woolly adelgids. Cluster along stems, on buds and on leaves; suck sap from plant; cause distorted or stunted growth; sticky honeydew forms on plant surfaces and encourages sooty mold growth.

What to do: squish small colonies by hand; dislodge with brisk water spray; encourage predatory insects and birds that feed on aphids; spray serious infestations with insecticidal soap or neem oil according to package directions.

Aster Yellows

see Viruses

Beetles

Many types and sizes; usually rounded in shape with hard, shell-like outer wings covering membranous inner wings. Some are beneficial, e.g., ladybird beetles ("ladybugs"); others, e.g., leaf skeletonizers and weevils, eat plants. Larvae: see Borers, Grubs. Leave wide range of chewing damage: make small or large holes in or around margins of leaves; consume entire leaves or areas between leaf veins ("skeletonize"); may also chew holes in flowers. Some bark beetle species carry deadly plant diseases.

What to do: pick beetles off at night and drop them into a container half filled with soapy water (soap prevents them from floating and climbing out).

Borer damage on squash

Mealybugs (see p. 44)

Blight

Fungal diseases. Many types; e.g., leaf blight, needle blight, snow blight. Leaves, stems and flowers blacken, rot and die.

What to do: thin stems to improve air circulation; keep mulch away from base of plants; remove debris from garden at end of growing season; remove and destroy infected plant parts.

Borers

Larvae of some moths, wasps and beetles. Worm-like; vary in size and get bigger as they bore through plants; among the most damaging plant pests. Burrow into plant stems, branches, leaves and/or roots; destroy vascular tissue (plant veins and arteries) and structural strength; weaken stems to cause breakage; leaves will wilt; may see tunnels in leaves, stems or roots; rhizomes may be hollowed out entirely or in part.

What to do: may be able to squish borers within leaves; remove and destroy bored parts; may need to dig up and destroy infected roots and rhizomes.

Bugs (True Bugs)

Green, brown, black or brightly colored and patterned, small insects, up to ½" long. Many are beneficial; a few pests, such as lace bugs, pierce plants to suck out sap; toxins may be injected that deform plants; sunken areas remain where pierced; leaves rip as they grow;

Lygus bug on cosmos flower

leaves, buds and new growth may be dwarfed and deformed.

What to do: remove debris and weeds from around plants in fall to destroy overwintering sites. Spray plants with insecticidal soap or neem oil according to package directions.

Case Bearers

see Caterpillars

Caterpillars

Larvae of butterflies, moths, sawflies; include bagworms, budworms, case bearers, cutworms, leaf rollers, leaf tiers and loopers. Chew foliage and buds; can completely defoliate a plant if infestation is severe.

What to do: removal from plant is best control; use high-pressure water and soap, or pick caterpillars off small plants by hand. Control biologically using the naturally occurring soil bacterium *Bacillus thuringiensis* var. *kurstaki* or *B.t.k.* (commercially available), which breaks down the gut lining of caterpillars.

Caterpillar on roses

Spittlebugs are unsightly but rarely cause damage.

Cutworms

see Caterpillars

Deer

Can decimate crops, woodlands and gardens; can kill saplings by rubbing their antlers on the trees, girdling the bark or snapping the trees in two; host ticks that carry Lyme disease.

Deer-chewed cedars

What to do: many deterrents work for a while: encircle immature shrubs with tall, upright sticks; place a fence around edibles (feed stores now sell inexpensive plastic webbing for deer control fences); use noisemaking devices or water spritzers to startle deer; mount flashy aluminum or moving devices throughout the garden.

Galls

Unusual swellings of plant tissues that may be caused by insects or diseases. Can affect leaves, buds, stems, flowers or fruit; often a specific gall affects a single genus or species.

What to do: cut galls out of plant and destroy them. Galls caused by insects usually contain the insect's eggs and juvenile stages; prevent such galls by controlling the insect before it lays eggs; otherwise try to remove and destroy infected tissue before young insects emerge. Insect galls are generally more unsightly than damaging to plants; galls caused by diseases often require destruction of plant. Don't place other plants susceptible to same disease in that location.

Leaf miners

Gray Mold

see Blight

Grubs

Larvae of different beetles; white or gray body; head may be white, gray, brown or reddish; usually curled in C-shape; commonly found below soil level. Problematic in lawns; may feed on roots of perennials; plant wilts despite regular watering; may pull easily out of ground in severe cases.

What to do: toss grubs onto a stone path, driveway, road or patio for birds to devour; apply parasitic nematodes or milky spore to infested soil (ask at your local garden center).

Leafhoppers and Treehoppers

Small, wedge-shaped insects; can be green, brown, gray or multi-colored; jump around frantically when disturbed. Suck juice from plant leaves; cause distorted growth; carry diseases such as aster yellows.

What to do: encourage predators by planting nectar-producing species such as coriander. Wash insects off with strong spray of water; spray with insecticidal soap or neem oil according to package directions.

Leaf Miners

Tiny, stubby larvae of some butterflies and moths; may be yellow or green. Tunnel within leaves leaving winding trails; tunneled areas lighter in color than rest of leaf; unsightly rather than major health risk to plant.

What to do: remove debris from area in fall to destroy overwintering sites; attract parasitic wasps with nectar plants such as yarrow and coriander; remove and destroy infected foliage; can sometimes squish larvae by hand within leaf.

Leaf Rollers

see Caterpillars

Leaf Skeletonizers

see Beetles

Leaf Spot

Two common types: one caused by bacteria and the other by fungi. Bacterial: small, brown or purple spots grow to encompass entire leaves; leaves may drop. Fungal: black, brown or yellow spots; leaves wither; e.g., scab, tar spot.

What to do: Bacterial: infection more severe; must remove entire plant. Fungal: remove and destroy infected plant parts; sterilize removal tools; avoid wetting foliage or touching wet foliage; remove and destroy debris at end of growing season.

Cabbage root maggots on radishes

Maggots

Larvae of several species of flies (cabbage root maggots, carrot rust flies). Small, white or gray, worm-like. Tunnel into roots of a variety of plants, including many root vegetables. Stunt plants and disfigure roots; serious infestations can kill plants.

What to do: use floating row covers to prevent flies from laying eggs near roots. Apply parasitic nematodes to soil around plants. Use an early crop of radishes as a trap crop. Pull them up and destroy them as soon as they become infested with maggots.

Mealybugs

Tiny, crawling insects related to aphids; appear to be covered with white fuzz or flour. Sucking damage stunts and stresses plant; excrete honeydew that promotes growth of sooty mold.

What to do: remove by hand from smaller plants; wash plant with soap and water or wipe with alcohol-soaked swabs; remove heavily infested leaves; encourage or introduce natural predators such as mealybug destroyer beetle and parasitic wasps; spray with insecticidal soap. Larvae of mealybug destroyer beetles look like very large mealybugs.

Mice

Burrow under mulch in winter, chewing plant roots, bark, tulip bulbs and many other underground goodies; even plants or roots stored in cool porches, garages or sheds are fair game.

What to do: fine wire mesh can prevent mice from getting at your plants in winter, though the rodents are quite ingenious and may find their way through or around any barrier you erect; bulbs and lifted roots can be rolled in talcum powder, garlic powder or bulb protectant spray before storing or planting. Get a cat, or borrow your neighbor's, if you must.

Mildew

Two types, both caused by fungus, but with slightly different symptoms. Downy mildew: yellow spots on upper sides of leaves and downy fuzz on undersides; fuzz may be yellow, white or gray. Powdery mildew: white or gray, powdery coating on leaf surfaces that doesn't brush off.

What to do: choose resistant cultivars; space plants well; thin stems to encourage air circulation; tidy any debris in fall; remove and destroy infected parts.

Mites

Tiny, red, yellow or green, eight-legged relatives of spiders, e.g., bud mites, spider mites, spruce mites; almost invisible to naked eye; do not eat insects but may spin webs; usually found on undersides of plant leaves; may be fine webbing on leaves and stems or mites moving on leaf undersides. Suck juice out of leaves; leaves become discolored and speckled, then turn brown and shrivel up.

What to do: wash off with strong spray of water daily until all signs of infestation are gone; predatory mites are available through garden centers; spray plants with insecticidal soap.

Moles and Gophers

Burrow under the soil, tunneling throughout your property in search of insects, grubs and earthworms; tunnels can create runways for voles that will eat your plants from below ground.

What to do: castor oil (the primary ingredient in most repellents made to thwart moles and gophers) spilled down the mole's runway is effective and is

Powdery mildew on black-eyed Susan

available in granulated pellet form, too; noisemakers and predator urine are also useful; humane trapping is effective, as is having a cat or dog.

Mosaic

see Viruses

Nematodes

Tiny worms that give plants disease symptoms; one type infects foliage and stems; the other infects roots. Foliar: leaves have yellow spots that turn brown; leaves shrivel and wither; problem starts on lowest leaves and works up plant. Root-knot: plant is stunted and may wilt; yellow spots on leaves; roots have tiny bumps or knots.

What to do: mulch soil; add organic matter; clean up debris in fall; don't touch wet foliage of infected plants; add parasitic nematodes to soil; remove infected plants in extreme cases.

Rabbits

Can eat as much of your garden as deer and munch on the bark of trees and shrubs.

What to do: deterrents that work for deer usually keep rabbits away; red pepper sprinkled on the plants can work as well, as will humane trapping; having a cat or dog to patrol your garden may also be effective.

Raccoons

Are especially fond of fruit and some vegetables; can carry rabies and canine distemper; also eat grubs, insects and mice, so can sometimes be helpful to gardeners.

What to do: don't allow access to garbage or pet food; humane traps and relocation are best solutions; call your local SPCA or Humane Society to relocate individuals.

Rot

Several different fungi that affect different parts of plant, sometimes even killing it. Crown rot: affects base of plant, causing stems to blacken and fall over and leaves to yellow and wilt. Root rot: leaves turn yellow and plant wilts; digging up plant shows roots rotted away. White rot: a "watery decay fungus" that affects any part of plant; cell walls appear to break down, releasing fluids.

What to do: keep soil well drained; don't damage plant if you are digging around it; keep mulches away from plant base; destroy infected plant if whole plant is affected.

Rust

Fungi; pale spots on upper leaf surfaces; orange, fuzzy or dusty spots on leaf undersides; e.g., blister rust, hollyhock rust.

What to do: choose rust-resistant varieties and cultivars; avoid handling wet leaves; provide plant with good air circulation; clear up garden debris at end of growing season; remove and destroy infected plant parts.

Scab

see Leaf Spot

Scale Insects

Tiny, shelled insects that suck sap, weakening and possibly killing plant or making it vulnerable to other problems; once female scale insect has pierced plant with mouthpart, it is there for life; juvenile scale insects are called crawlers.

What to do: use alcohol-soaked swabs to wipe bugs off; spray plant with water to dislodge crawlers; prune out heavily infested branches; encourage natural predators and parasites; spray dormant oil in spring before bud break.

Slugs and Snails

Slugs lack shells; snails have a spiral shell; both have slimy, smooth skin; can be up to 8" long but are often much smaller; gray, green, black, beige, yellow or spotted. Leave large, ragged holes in leaves and silvery slime trails on and around plants.

What to do: attach strips of copper to wood around raised beds or smaller boards inserted around susceptible groups of plants (slugs and snails get shocked if they touch copper surfaces); pick off by hand in the evening and squish or drop in a can of soapy water; spread wood ash or diatomaceous earth (available in garden centers) around plants (it pierces their soft bodies and causes them to dehydrate); slug baits containing iron phosphate are not harmful to humans or animals and control slugs very well; you can also set small clay flower pots on their sides, stuffed with damp newspaper, in the vegetable garden to lure the slugs. If slugs damaged garden last season, begin controls as soon as new green shoots appear in spring.

Sooty Mold

Fungus; thin black film forms on leaf surfaces and reduces amount of light getting to leaf surfaces.

Snails (top) and slugs (center) can cause a lot of chewing damage to plants (bottom).

What to do: wipe mold off leaf surfaces; control aphids, mealybugs and whiteflies (honeydew left on leaves encourages mold).

Squirrels

Unearth and eat bulbs and corms, as well as flowers, fruits and vegetables; chew on sugar maples and hone their other teeth on almost everything else; raid birdfeeders and often eat the feeder itself; bury their food for later consumption, which can result in seeds germinating and plants springing up where you never wanted them.

Sticky pheremone traps are useful for monitoring insect populations.

What to do: sprinkle red cayenne pepper on ground after planting bulbs; cut heavy metal screening (hardware cloth) to fit around the plant stem; caging entire plants is effective if you don't mind your garden looking like a zoo; removing enticing food supplies is effective, but often impractical; trapping and moving is one option but usually results in other squirrels moving in to take their place.

Tar Spot

see Leaf Spot

Thrips

Tiny, slender, yellow, black or brown insects; narrow, fringed wings; difficult to see but may be visible if you disturb them by blowing gently on an infested flower. Suck juice out of plant cells, particularly in flowers and buds, causing mottled petals and leaves, dying buds, distorted and stunted growth.

What to do: remove and destroy infected plant parts; encourage native predatory insects with nectar plants such as yarrow and coriander; spray severe infestations with insecticidal soap or neem oil according to package directions.

Viruses

Include aster yellows, mosaic virus and ringspot virus. Plant may be stunted and leaves and flowers distorted, streaked or discolored.

What to do: viral diseases in plants cannot be treated. Control disease-spreading insects, such as aphids, leafhoppers and whiteflies; destroy infected plants.

Voles

Mouse-like creatures that damage plants at or just beneath the soil surface; mostly herbivorous, feeding on a variety of grasses, vegetables, herbaceous plants, bulbs (lilies are a favorite) and tubers; also eat bark and roots of trees, usually in fall or winter, and store seeds and other plant matter in underground chambers.

What to do: wire fences at least 12" high with a mesh size of ½" or less and buried 6–8" deep can help exclude voles from gardens. These fences can either stand alone or be attached to the bottom of an existing fence. A weed-free barrier on the outside of the fence will increase its effectiveness. Burrow fumigants do not effectively control voles because the vole's burrow system is shallow and has many open holes. Electromagnetic or ultrasonic devices and flooding are also ineffective. When vole populations are not numerous or concentrated in a small area, trapping may be effective. Use enough traps to control the population: for a small garden, use at least 12 traps, and for larger areas, 50 or more may be needed. Again, a dog or cat is a deterrent. Do not use poisonous repellents or baits if your pets or children romp around the garden.

Weevils

see Beetles

Whiteflies

Tiny, white, moth-like flying insects that flutter up into the air when plant is disturbed; live on undersides of plant leaves. Suck juice out of leaves, causing yellowed leaves and weakened plants; leave behind sticky honeydew on leaves, encouraging sooty mold growth.

What to do: destroy weeds where insects may live; attract native predatory beetles and parasitic wasps with nectar plants such as yarrow and coriander; spray severe cases with insecticidal soap; make a sticky flypaper-like trap by mounting a tin can on a stake, wrapping can with yellow paper and covering it with a clear plastic bag smeared with petroleum jelly (replace bag when it's covered in flies). Don't let plants dry out.

Wilt

If watering doesn't help wilted plants, one of two wilt fungi may be to blame. *Fusarium* wilt: plant wilts; leaves turn yellow then die; symptoms generally appear first on one part of plant before spreading to other parts. *Verticillium* wilt: plant wilts; leaves curl up at edges; leaves turn yellow then drop off; plant may die.

What to do: both wilts are difficult to control; choose resistant plant varieties and cultivars; clean up debris at end of growing season; destroy infected plants; solarize (sterilize) soil before replanting (may help if entire bed of plants is lost to these fungi)—contact local garden center for assistance.

Woolly Adelgids

see Aphids

Worms

see Caterpillars, Nematodes

Homemade Insecticidal Soap

1 tsp mild dish detergent or pure soap (biodegradable options are available)

4 cups water

Mix in a clean spray bottle and spray the surfaces of your plants. Rinse well within an hour to avoid foliage discoloration.

About this Guide

The plants featured here are organized alphabetically by their most common familiar names. Additional common names appear as well. This system enables you to find a plant easily if you are familiar with only the common name. The scientific or botanical name is always listed after the common name. We encourage you to learn these botanical names. Several plants may share the same common name, and common names vary from region to region. Only the botanical name identifies the specific plant anywhere in the world.

Clearly indicated within each entry are the plant's height and spread ranges, outstanding features and hardiness zone(s), if applicable. Each entry

gives clear instructions for starting and growing the plants, and recommends many favorite selections. If height and spread ranges or hardiness zones are not given for every recommended plant, assume these ranges are the same as those provided in the Features section. Your local garden center will have any additional information about the plant and will help you make your plant selections.

Finally, the Problems and Pests section in each account deals with issues that afflict your garden plants from time to time. More information about common pests and diseases, including prevention and treatment, is given in the Problems and Pests section in the Introduction.

Arbutus

Strawberry Tree, Madrone

Arbutus

Features: spreading or shrubby, broad-leaved evergreen tree; flowers; edible fruit Height: 5–70' Spread: 5–70'

Birds love arbutus berries, but people can suffer stomachaches if they eat too many of these strawberry-like fruit. The trees grew wild on the island where I grew up, and they were perfect for building tree forts.

Starting

Purchase young trees from nursery. Buy in fall to taste the fruit, as flavor varies from tree to tree and most of the time the fruit is bland.

Growing

Plant arbutus in a **sheltered** spot in **full sun**. The soil should be **fertile**, **humus rich** and **well drained**. Do not overwater—these trees are drought and salt tolerant. They do not like having their roots disturbed, so pick a spot and leave the tree there. They do not usually need pruning.

Harvesting

Fruit is yellow at first then turns red in fall when it matures. A mealy and bland taste is usual, but individual trees sometimes have sweet fruit.

Tips

Use in a woodland garden or as a specimen tree. Smaller cultivars work in a shrub or mixed border.

Recommended

***A.* 'Mariana'** has large, green leaves, rosy pink flowers and red and yellow fruit. It grows 20–30' tall.

A. menziesii (arbutus, Pacific madrone) is a spreading or shrubby tree native to western North America. It grows 50–70' tall and wide. White flowers appear in erect clusters in early summer, followed by the warty, red fruits, borne in equally striking clusters that can stay on the tree until December.

A. unedo (Mediterranean strawberry tree) is a spreading, shrubby tree with shredding, exfoliating red-brown bark. It grows 15–30' tall and wide. White flowers appear in fall, followed by warty, red fruits. **'Compacta'** is a slow-growing tree with slightly contorted branches. It grows to about 15' tall. **'Elfin King'** is a compact, bushy form that flowers and fruits profusely. It grows 5–10' tall and wide.

A. menziesii bark (above)

Problems and Pests

Fungal leaf spot, tent caterpillars and scale insects can cause some trouble.

Both arbutus *(Latin) and* madroño *(Spanish) mean "strawberry tree," in reference to the bright red fruits of these magnificent evergreen trees.*

Artichokes

Cynara

Features: bushy, tender perennial; spiny foliage; edible flower buds; purple, thistle-like flowers **Height:** 2–7' **Spread:** 2–4'

Artichokes are one of those vegetables that require a lot of extra effort to grow successfully in western Washington and Oregon. These members of the thistle family are tender perennials native to the Mediterranean. With a thick mulch, they may survive winter in the milder parts of the region. For the rest of us, choosing a quick-maturing cultivar and starting very early is the best way to enjoy homegrown artichokes.

Starting

Start seed indoors about eight weeks before the last frost date or up to 12 weeks before the last frost date if you have a very bright location or supplemental lighting available. Plant seeds into the largest size of peat pots. These plants don't like to have their roots disturbed, and you want to encourage them to grow as much as possible before they are planted into the garden. You can also purchase started plants, but you may have to shop early because their availability is low, and often only specialty nurseries carry them.

Plant artichokes outdoors after the last frost date. These are very wide-spreading plants, so plant them at least 2' apart the first year and eventually up to 4' apart, if you overwinter them successfully.

Growing

Artichokes grow best in **full sun**, though they appreciate some afternoon shade in hotter locations. The soil should be **fertile, humus rich, moist** and **well drained**. Mix compost into the soil in spring, and use a layer of compost on the soil as a mulch. These plants like plenty of water but

don't like a soggy soil. Replenish the compost mulch in mid-summer, if needed. If the buds are just beginning to form and heavy frost is expected, you can cover the plants with sheets at night to extend the growing season.

Near Puget Sound or in the milder parts of the region, artichokes may survive winter. If they are planted near a house foundation or other place where the soil doesn't freeze completely, plants can be protected with a thick layer of straw. A layer 12–18" deep will help keep the soil temperature more consistent.

In other parts of the region, we can either treat these plants like annuals or dig up the roots once the leaves die back in fall, clean off any soil and store them in slightly moistened peat or sphagnum moss in a cold, frost-free location for winter. Check them over winter, and if they are beginning to sprout, pot them and keep them in the brightest indoor location you can find until they can be planted outdoors again.

Harvesting

Artichokes produce one large flower bud on the central stalk and many smaller flower buds on the side shoots. The flower buds are rounded and made up of tightly packed scales. They are usually ready for harvest as the scales just begin to loosen. With a sharp knife, cut the flower bud from the plant about

4–6" below the bud's base. With luck and an early enough start, you should have a good crop the first year in fall. If overwintered successfully, artichokes generally flower earlier in the season, in June or July.

Tips

Artichoke plants make dramatic additions to vegetable and ornamental gardens. The flowers are quite stunning and can be used in fresh flower arrangements if not eaten while in bud form. If you are going to try to overwinter artichokes in the ground, plant them in a place where the soil stays warmest in winter. This spot is often dry in summer, so mulch well to retain soil moisture.

Recommended

C. scolymus forms a large clump of deeply lobed, pointy-tipped, gray-green leaves. It grows 2–7' tall and spreads 2–4'. In fall or sometimes summer, it bears large, scaled flower buds that open if not picked for eating. **'Green Globe'** is a popular cultivar because it is tasty and quick to mature and flower. **'Imperial Star'** is a good choice for cool summer areas because it is grown as an annual, producing artichokes only 90 days after transplanting.

Problems and Pests

Rare problems with mold, root rot, slugs and aphids are possible.

Asparagus

Asparagus

Features: perennial; edible spring shoots; ferny growth; small, white, summer flowers; decorative red fruit **Height:** 2–5' **Spread:** 2–4'

The large, ferny growth that asparagus develops comes as quite a surprise to first-time growers who may have seen only tidy bunches of spears at the grocery store. Asparagus is a member of the Lily family, and well-established plants can last a lifetime, producing tasty spears every spring.

Starting

Asparagus can be started from seed, or you can buy roots (also called crowns). A plant started from seed will be ready to start harvesting the third spring. A plant started from roots will be ready to begin harvesting the second spring.

Plant purchased roots into a well-prepared area. Work plenty of compost into the bed, then dig a trench or hole about 18" deep. Lay the roots 18–24" apart from each other and other plants. Cover the roots with 2–4" of soil and, as they sprout up, gradually cover them with more soil until the trench or hole is filled. Water and mulch well.

Plant seeds indoors in flats or peat pots about six to eight weeks before you will be planting them into the garden. Use larger pots if the seedlings get too big before the last frost date has passed and you can move them outside. The first year, you should plant the seedlings at the soil level they are at in their pots. Keep them well watered and mulch them with compost. The second summer, they can be planted as for roots, described above.

Asparagus is a good companion plant for tomatoes. Tomato plants repel the asparagus beetle, and asparagus repels root nematodes that can harm tomato plants.

A. officinalis 'Mary Washington' (above)

Growing

Asparagus grows well in **full sun** or **partial shade** with protection from the hot afternoon sun. The soil should be **fertile, humus rich, moist** and **well drained**. Apply a 4" layer of compost in spring and late summer. Weed regularly because this plant is most productive if it doesn't have a lot of competition from other plants.

Harvesting

As mentioned, asparagus spears that were started from roots are ready to be harvested two years after planting; spears started from seeds are ready in three years. Snap or cut the spears off at ground level for up to about four weeks in spring and early summer. When new spears are thinner than a pencil, you should stop harvesting and let the plants grow in. Add a new layer of compost to the soil when you have finished harvesting.

Tips

This hardy perennial plant is a welcome treat in spring and a beautiful addition to the back of a border.

Recommended

A. officinalis forms an airy mound of ferny growth. It grows 2–5' tall and spreads 2–4'. Small, white, summer flowers are followed by bright red berries, which can be collected for starting new plants. **'Jersey Giant'** and **'Jersey Knight'** are two all male varieties that bear larger spikes, but **'Mary Washington'** is the traditional variety that bears both male and female sprouts.

Problems and Pests

Rust can be a problem, so choose resistant cultivars.

A. officinalis 'Mary Washington' (above)

Asparagus has been a cultivated vegetable crop for over 2000 years.

Asparagus is dioecious—male and female flowers are borne on separate plants. Male plants are reputed to produce the greatest number of spears.

Basil

Ocimum

Features: fragrant, decorative leaves; white or light purple flowers
Height: 12–24" **Spread:** 12–18"

The sweet, fragrant leaves of basil add a delicious, licorice-like flavor to salads and tomato-based dishes.

Starting

Basil is easy to start from seed. Start seeds indoors about four weeks before the last frost date, or sow seed directly outdoors once the last frost date has passed and the soil has warmed up. Press seeds into the soil and sift a little more soil over them. Keep the soil moist. Generally seeds will germinate within a week.

Growing

Basil grows best in a **warm, sheltered** location in **full sun**. Do not place plants outdoors too early. Wait until June, when the night temperatures are warmer. The soil should be **fertile, moist** and **well drained**. Pinch the flowering tips regularly to encourage bushy growth and leaf production. Mulch soil to retain water and to keep weeds down.

Harvesting

Pluck leaves or pinch back stem tips as needed. Basil is tastiest if used fresh, but it can be dried or frozen for use in winter.

Tips

Although basil grows best in a warm spot outdoors, it can be grown successfully indoors in a pot by a bright window, providing you with fresh leaves all year.

O. basilicum

Recommended

O. basilicum is a bushy annual or short-lived perennial with bright green, fragrant leaves. It bears white or light purple flowers in mid- to late summer. There are many cultivars of basil, with varied leaf sizes, shapes, colors and flavors. The purple Thai basil is among the most decorative and looks great in a pot on a sunny patio.

Problems and Pests

Fusarium wilt is probably the worst problem that afflicts basil.

Basil makes a great companion for tomatoes because they both require warm, moist growing conditions and are delicious when eaten together.

Bay Laurel

Laurus

Features: tender, evergreen shrub; aromatic foliage
Height: 1–4' **Spread:** 8–24"

Bay leaves are commonly used in soups and stews. The plants are easy to grow and are attractive houseplants.

Starting

Bay laurel can be started from seed, but germination may take up to six months. Plant the seeds in warm soil and keep the environment warm and moist, but not wet or the seeds will rot. It is simpler to purchase started plants, which are available from specialty growers and nurseries.

Growing

Bay laurel grows well in **full sun** or **partial shade**. A plant that will be moved indoors for winter should be grown in partial or light shade in summer. The soil should be **fertile, moist** and **well drained**. This plant is shallow-rooted and can dry out quickly in hot or windy weather. Mulch to reduce evaporation.

In areas where winters are cold, move bay laurel indoors; in protected areas near the coast, your bay laurel will survive the winter outdoors. As a houseplant, keep it in a cool but sunny room; continue to water it regularly, as needed.

Harvesting

It may take a couple of years before your bay laurel is leafy enough to be used on a regular basis. Pick fresh leaves as needed to use in cooking. Leaves can be dried and stored for later use, but this plant is evergreen, so you should be able to pick fresh leaves all year.

Tips

Bay laurel makes an attractive addition to patios, decks and the steps of a staircase with other potted herbs, vegetables and flowers.

L. nobilis

Recommended

L. nobilis is a bushy, evergreen shrub. It grows up to 40' tall in the Mediterranean, where it is native. In containers, it can be kept to a far more manageable size of 1–4' in height and 8–24" in spread. **'Aureus'** is a cultivar with golden yellow foliage.

Problems and Pests

Rare problems with scale insects and mealybugs can occur, but the pests are usually easy enough to wash or rub off smaller plants. Powdery mildew can occur in poorly ventilated situations.

Beans

Phaseolus

Features: bushy or twining, tender annual; attractive foliage; red, white or bicolored flowers; edible pods or seeds **Height:** 1–8' **Spread:** 12–18"

This incredibly diverse group of legumes includes beans eaten as pods, as immature seeds or as mature, dry seeds. Plants can be low and bushy or tall and twining. Plants are often prolific, and some are very attractive.

Starting

Beans are one of the easiest plants to grow from seed. They are large and easy to handle, and they sprout quickly in warm, moist soil. Plant them directly in the garden after the last frost date has passed and the soil has warmed up—usually in June or even as late as July. They can be planted 4–8" apart.

Growing

Beans grow best in **full sun**, but they tolerate some light afternoon shade. The soil should be of **average fertility** and **well drained**. Bush beans are self-supporting, but climbing beans need a pole or trellis to grow up. The support structure should be in place at planting time to avoid disturbing the young plants or damaging their roots.

Bush beans can become less productive and look unattractive as summer wears on. Pull them up and plant something else in their place, or plant them with companions that mature more slowly to fill in the space left by the faded bean plants.

Harvesting

The most important thing to remember when harvesting beans is to do so only when the foliage is dry. Touching wet foliage or plants encourages the spread of disease.

Different types of beans should be picked at different stages in their development. Green, runner, wax or snap beans are picked once the pod is a good size but still young and tender. As they mature, they become stringy, woody and dry.

Legumes, such as beans, are known for being able to fix nitrogen from the air into the soil. They do this through a symbiotic relationship with bacteria, which attach to the roots as small nodules. The bacteria turn the nitrogen from the air into usable nitrogen for the plant; in return, the plant feeds and supports the bacteria. The bacteria are present in most soils and are also available for purchase as a soil innoculant. Some bean seeds are also pre-treated with the bacteria.

P. vulgaris (above), P. lunatus (below)

Beans that are eaten as immature seeds should be picked when the pods are full and the seeds are fleshy and moist.

Beans for drying are left to mature on the plant. Once the plant begins to die back and before the seedpods open, the entire plant is cut off at ground level and hung upside down to finish drying. The beans can then be removed from the pods and stored in airtight containers.

Tips

Beans are very ornamental, with attractive leaves and plentiful flowers. Climbing beans can be grown up fences, trellises, obelisks and poles to create a screen or feature planting. Bush beans can be used to make low, temporary hedges or can be planted in small groups in a border.

Recommended

P. coccineus (runner bean) is a vigorous climbing plant, with red or sometimes white or bicolored flowers. **'Scarlet Runner'** has bright red flowers and is one of the best and best-known cultivars. The beans can be eaten with the pod when they are young and tender, or they can be left to mature, and the pink and purple spotted beans can be dried. Plants produce edible beans in 70 days but need about 100 days for dry beans.

P. lunatus (lima bean) may be climbing or bush, depending on the cultivar. The beans are eaten as immature seeds and should be picked when the pods are plump, but the seeds are still tender. They take 70 to 85 days to mature. **'Fordhook'** is a popular bush variety, and

'King of the Garden' is a good climbing selection.

P. vulgaris (wax bean, green bean, bush bean, snap bean, dry bean) is probably the largest group of beans and includes bush beans and pole beans. Some are eaten immature in the pod, and others are grown to maturity and used as dry beans. Bush bean cultivars may be yellow, such as **'Carson'** and **'Eureka'**; green, such as **'Provider'** and **'Jade'**; or purple-podded, such as **'Royal Burgundy.'** Purple beans turn bright green when cooked. Bush beans take 50 to 60 days to mature. Pole beans, such as **'Blue Lake'** and **'Kentucky Blue,'** take 50 to 55 days to mature. Dry beans are usually bush plants and take about 100 days to mature. They include kidney, pinto and navy beans. West of the Cascades, fava beans are a popular variety and do well in cool weather.

Problems and Pests

Problems with leaf spot, bacterial blight, rust, bean beetles and aphids can occur. Disease-infected plants should be destroyed, not composted, once you've harvested what you can.

Climbing beans are popular among gardeners with limited space because you can get more beans for less space. Pole beans are especially sensitive to windy weather.

P. coccineus 'Scarlet Runner' (above & below)

Beets

Beta

Features: clump-forming biennial grown as an annual; attractive leaves; red, yellow or red-and-white-ringed, edible root **Height:** 8–18" **Spread:** 4–12"

Beets are versatile vegetables. The plump, rounded or cylindrical roots are most commonly eaten. The tops are also edible and can be compared in flavor to spinach and Swiss chard. Beets and Swiss chard are closely related, both being members of the genus *Beta*.

B. vulgaris (above & below)

Starting

The corky, wrinkled seed of the beet is actually a dry fruit that contains several tiny seeds. Plant it directly in the garden around the last frost date. Even if you space the seeds 3–6" apart, you will probably have to thin a bit because several plants can sprout from each fruit. Beets are fairly quick to mature, and a second crop can often be planted in mid-summer for a fall crop; then west of the Cascades a mulch can keep your summer-planted beets fresh all winter.

Growing

Beets grow well in **full sun** or **partial shade**. They grow best in cool weather. The soil should be **fertile, moist** and **well drained**. Mulch lightly with compost to maintain moisture and improve soil texture.

Harvesting

Beets mature in 45 to 80 days, depending on the variety. Short-season beets are best for immediate eating and preserving, and long-season beets are the better choice for storing.

Pick beets as soon as they are big enough to eat. They are tender when

young but can become woody as they mature.

You can pick leaves from the beets without pulling up the entire beet if you want to use the leaves for fresh or steamed greens. Don't pull all the leaves off a beet; just remove a few at a time from any one plant.

Tips

Beets have attractive red-veined, dark green foliage. These plants look good when planted in small groups in a border, and they make interesting edging plants. They can even be included in large mixed container plantings.

Recommended

B. vulgaris forms a dense rosette of glossy, dark green leaves, often with deep red stems and veins. It grows 8–18" tall and spreads 4–8". There are many cultivars available. **'Chioggia'** is an heirloom cultivar that produces red-and-white-ringed roots. **'Early Wonder'** and **'Red Ace'** are good red, round cultivars. **'Winter Keeper'** is a cold hardy variety that can be planted in midsummer west of the Cascades and will overwinter until spring when the beets can be harvested. The beautiful **'Golden'** variety adds color to stews and beet salads but does better in the dry summer areas east of the Cascades; it will struggle in cool, wet soil.

Problems and Pests

Beets are generally problem free, but occasional trouble with scab, root maggots and flea beetles can occur. To avoid flea beetles, plant your beets later in the season—after June 15.

Never fear if you get beet juice on your clothing; it won't stain. Dyers have been unsuccessfully trying to find a fixative for beet juice for centuries. Chemists inform us that the red molecule in the beet is very large and doesn't adhere to other molecules, so a fixative is unlikely to ever be found.

Blueberries

Vaccinium

Features: deciduous shrub; small, bell-shaped, white or pink flowers; edible fruit; attractive habit; bright red fall color **Height:** 4"–5' **Spread:** 1–5'

These attractive bushes are low and spreading or rounded and upright. The leaves turn a beautiful shade of red in fall. The plants are an admirable addition to any border, with the added bonus of delicious summer fruit. Blueberries are related to huckleberries and cranberries.

Starting

Plants can be purchased and planted at any time as long as the ground is workable. The best selection is generally in spring.

Growing

Blueberries grow well in **full sun, partial shade** or **light shade**. The soil should be of **average fertility, acidic, moist** and **well drained**. Blueberries grow best in areas where the soil is acidic and peaty or sandy. They do especially well near the coast, where high rainfall makes for the acidic soil they love.

Harvesting

Blueberries are ready for harvesting when they turn, not surprisingly, blue. Test one, and if it is sweet and tastes the way you expect, they are ready for harvest.

Tips

If you have naturally acidic soil, blueberries make an excellent choice for a fruit-bearing shrub in

V. angustifolium

a woody or mixed border. They make great edible hedges, and some varieties have spectacular fall foliage.

Recommended

V. angustifolium* var. *laevifolium (lowbush blueberry, wild blueberry) is a low, bushy, spreading shrub with small, glossy, green leaves that turn red in fall. It grows 4–24" tall and spreads 12–24". Clusters of small, bell-shaped, white or pink flowers are produced in spring, followed by small, round fruit that ripens to dark blue in mid-summer. (Zones 2–8)

V. corymbosum (highbush blueberry) is a bushy, upright, arching shrub with green leaves that turn red or yellow in fall. It grows 3–5' tall with an equal spread. Clusters of white or pink flowers at the ends of the branches in spring are followed by berries that ripen to bright blue in summer. Several cultivars are available, including **'Bluecrop,'** with tart, light blue berries; **'Olympia,'**

developed in Olympia, Washington; and **'Darrow,'** a heavy producer with very large fruit. (Zones 3–8)

Problems and Pests

Rare problems with caterpillars, rust, scale, powdery mildew and root rot can occur. To keep the birds from eating your berries, drape the shrubs with tulle netting just before the berries ripen.

A handy way to preserve blueberries is to spread them on a cookie sheet and put them in a freezer. Once they are frozen, they can be put into an airtight bag and kept in the freezer. The berries will be frozen individually, rather than in a solid block, making it easy to measure out just what you need for a single recipe or serving.

Borage

Borago

Features: bushy, bristly annual; edible, bristly leaves; edible, blue, purple or white, summer flowers **Height:** 18–28" **Spread:** 18–24"

Borage leaves and flowers are both edible, making an interesting addition to salads. They have a very light, cucumber-like flavor. The flowers can also be frozen in ice cubes or used to decorate cakes and other desserts.

Starting
Seed can be sown directly in the garden in spring. This plant resents being transplanted because it has long taproots, but it recovers fairly quickly if moved when young.

Growing
Borage grows well in **full sun** or **partial shade**. The soil should be of **average fertility, light** and **well drained**, but this plant adapts to most conditions. It makes a good choice in a dry location because it doesn't require much water to thrive.

Borage is a vigorous self-seeder. Once you have established it in your garden, you will never have to plant it again. Young seedlings can be pulled up if they are growing where you don't want them.

Harvesting
Pick borage leaves when they are young and fuzzy—they become rather bristly as they mature. Flowers can be picked any time after they open; they tend to change color from blue to pinkish mauve as they mature.

Tips
Borage makes an attractive addition to herb and vegetable gardens, as well as to flower beds and borders. The plant should be pinched back when it is young to encourage bushy growth; otherwise, it tends to flop over and develop a sprawling habit.

Recommended
B. officinalis is a bushy plant with bristly leaves and stems. It bears clusters of star-shaped, blue or purple flowers from mid-summer to fall. A white-flowered variety is available.

Problems and Pests
Rare outbreaks of powdery mildew and aphids are possible but don't seem to be detrimental to the plant.

Borage attracts bees, butterflies and other pollinators and beneficial insects to the garden.

Broccoli

Brassica

Features: bushy, upright annual; powdery, blue-green foliage; dense clusters of edible flowers **Height:** 12–36" **Spread:** 12–18"

Although usually thought of as a vegetable, broccoli could more accurately be called an edible flower. It is the large, dense flower clusters that are generally eaten, though the stems and leaves are also edible.

Starting

Broccoli can be started from seed indoors or planted directly into the garden. Sow seeds indoors four to six weeks before the last frost date, and plant seedlings out or direct sow into the garden around the last frost date.

Growing

Broccoli grows best in **full sun**. The soil should be **fertile, moist** and **well drained**. Broccoli performs best in cooler weather. Mix compost into the soil, and add a layer of mulch to keep the soil moist. Don't let this plant dry out excessively because it can delay flowering. You can plant broccoli in mid-summer for fall harvest in the mild winter areas near the coast.

Harvesting

Broccoli forms a central head (broccoli), and some varieties also produce side shoots. Pick the heads by cutting them cleanly from the plant with a sharp knife. If you leave them for too long on the plant, the bright yellow flowers will open.

Tips

Broccoli, with its blue-green foliage, is an interesting accent plant. Tuck it in groups of three or so into your borders and mixed beds for a striking contrast. This plant is susceptible to quite a few pest and disease problems, and spacing it out rather than planting it in rows helps reduce the severity of potential problems.

Choosing a side-shoot-producing plant versus a main-head-only variety is a matter of personal preference. If you have a large family or plan to freeze some florets, you may prefer a large-headed selection. If there are only a few people in the household, or you want to enjoy the broccoli for a longer time period without storing any, you may prefer the small-headed varieties that produce plenty of side shoots.

Recommended

B. oleracea* var. *botrytis is an upright plant with a stout, central, leafy stem. Flowers form at the top of the plant and sometimes from side shoots that emerge from just above each leaf. Plants grow 12–36" tall and spread 12–18". Maturity dates vary from 45 to 70 days.

'Calbrese' ('Calabria') produces plenty of side shoots. **'Early Dividend'** will yield in 50 days with lots of side shoots. **'Gypsy'** produces a large central head and is one of the most heat-tolerant cultivars. **'Packman'** produces generous yields and handles summer heat well. **'Premium Crop'** is an All America Selections winner that is early to mature and produces plenty of side shoots.

Cabbage white butterflies are common pests for all Brassicas, and their tiny, green caterpillar larvae can be tough to spot in a head of broccoli. Break heads into pieces and soak them in salted water for 10 or so minutes before cooking. This kills the larvae and causes them to float to the surface.

Problems and Pests

Problems with cutworms, leaf miners, caterpillars, root maggots, cabbage white butterflies, white rust, downy mildew and powdery mildew can occur. Avoid planting this plant in the same spot in successive years. You can control worms, loopers and root maggots by covering your broccoli plants with a floating row cover like Reemay in early spring.

Brussels Sprouts

Brassica

Features: bushy, upright plant, grown as an annual; fat, edible buds sprout from leaf bases along the stem; blue-green foliage **Height:** 24" **Spread:** 18"

Love them or hate them, Brussels sprouts are at the very least a garden curiosity. The plants develop a stout central stem, and the sprouts form on the stem at the base of each leaf. By fall, the display is unique and eye-catching.

Starting

Brussels sprouts can be purchased as small transplants in spring, or seeds can be started indoors about six weeks before you expect to transplant them into the garden. Sowing the seeds into peat pots or pellets makes it easy to transplant the seedlings into the garden.

Growing

Brussels sprouts grow well in **full sun**. The soil should be **fertile, moist** and **well drained**. Brussels sprouts need a fairly long growing season to produce sprouts of any appreciable size, so they should be planted out as early as possible. Regular moisture encourages them to mature quickly, so keep the soil well mulched. Once you see sprouts starting to form, you may wish to remove some of the stem leaves to give the sprouts more room to grow.

Harvesting

Pick sprouts as soon as they are large and plump, but before they begin to open. A light frost can improve the flavor of the sprouts. The entire plant can be pulled up, and if you remove

Nutrient-rich and fiber-packed, Brussels sprouts deserve a more prominent place on our plates. People who complain that sprouts are mushy or bitter have probably had ones that were poorly prepared. When picked after a light frost, steamed so that they are cooked just through to the center and served with butter, they are delicious. You can also roast them in the oven to bring out their sweetness.

Brussels sprouts are categorized into early-, mid- and late-season varieties. If you want a fairly regular harvest, choose plants from each category. If you consider them a late-fall treat, look for mid- or late-season varieties.

the roots, leaves and top of the plant, the sprouts can be stored on the stem in a cool place for up to four weeks. Be sure to keep an eye on them because they can go bad quickly. They can also be frozen for later use.

Tips

Brussels sprouts create a leafy backdrop for your flowering annuals and perennials all summer; then, just as the garden is fading, they create an interesting focal point as the plump little sprouts develop.

Recommended

B. oleracea* var. *gemnifera is an upright plant that develops a single leafy stem. Sprouts form at the base of each leaf along the stem. The leaves are blue-green, often with white mid-ribs and stems. **'Vancouver'** and **'Rubine'** are two varieties for cool summer areas. East of the Cascades, the variety **'Bubbles'** has excellent flavor and more heat resistance to hot summer weather.

Problems and Pests

Problems with cutworms, leaf miners, caterpillars, root maggots, aphids, cabbage white butterfly larvae, white rust, downy mildew and powdery mildew can occur.

To avoid overcooking Brussels sprouts, cut an X in the stem about one-quarter of the way through each sprout to help the inside cook at the same rate as the outside.

Cabbage

Brassica

Features: biennial grown as an annual; dense, round, red, purple, blue-green or green, leafy clumps; smooth or crinkled leaves **Height:** 18–24" **Spread:** 12–18"

Cabbages are easy to grow, and they create a dense, leafy, often colorful display. They come in three forms: green with smooth leaves, green with very crinkled leaves and red or purple with (usually) smooth leaves.

Starting
Start seeds indoors about four weeks before you plan to transplant them outdoors. You can also direct sow them into the garden around the last frost date as long as the soil has warmed up a bit.

Growing
Cabbages grow best in **full sun.** They prefer cool growing conditions and benefit from mulch to retain moisture during hot weather. The soil should be **fertile, moist** and **well drained**.

Harvesting
The leaves of young plants can be eaten. When a good-sized head has developed, cut it cleanly from the plant. Smaller heads often develop once the main head has been cut. Plants are frost hardy, and the last of the cabbages can be left in the ground through fall, then stored in a cold, frost-free location. Cabbages that take a long time to mature generally store better than quick-maturing types.

Tips
Choose a variety of cabbages because the different colors, textures and maturing times create a more interesting display, whether in rows or mixed into your borders.

Recommended
B. oleracea* var. *capitata is a low, leafy rosette that develops a dense head over summer. Leaves may be green, blue-green, red or purple and smooth or crinkled. It matures in 60 to 140 days, depending on the variety. Some popular smooth-leaved, green varieties are **'January King'** and **'Bartolo.' 'Derby Day'** is a cool-season cabbage that does well where spring is wet and chilly. Popular crinkled or savoy types include **'Alcosa,' 'Savoy King,' 'Samantha'** and **'Wirosa.'** Popular red cabbages include **'Ruby Ball,' 'Red Acre Early,' 'Red Drumhead,' 'Red Express'** and **'Ruby Perfection.'**

Problems and Pests
Problems with cutworms, leaf miners, caterpillars, root maggots, aphids, cabbage white butterfly larvae, white rust, downy mildew and powdery mildew can occur.

Don't forget that all members of the cabbage genus (Brassica) *are susceptible to many of the same pests and diseases. Don't plant them in the same spot two years in a row, particularly if you've had disease problems in that area. Like broccoli, cabbages can be protected from many pests simply by covering the new sprouts with Reemay or floating row covers. Spray plants with* B.t. (Bacillus thuriengiensis) *for heavy infestations.*

Calendula

Pot Marigold, English Marigold

Calendula

Features: hardy annual; yellow, orange, cream, gold or apricot, edible flowers; long blooming period **Height:** 10–24" **Spread:** 8–20"

Bright and charming, calendula produces attractive, colorful flowers in summer and fall.

Starting

Calendula grows easily from seed. Sprinkle the seeds where you want them to grow (they don't like to be transplanted), and cover them lightly with soil. They will sprout within a week.

Young plants are often difficult to find in nurseries because this plant is so quick and easy to start from seed.

Growing

Calendula does well in **full sun** or **partial shade**. The soil should be of **average fertility** and **well drained**. Calendula likes cool weather and can withstand a moderate frost.

Deadhead the plant to prolong blooming and keep it looking neat. If your plant fades in the summer heat, cut it back to 4–6" above the ground to encourage new growth. A fading plant can also be pulled up and new seeds planted. Both methods provide a good fall display.

Harvesting

Flowers can be picked as needed for fresh use, or the petals can be dried for later use.

Tips

This informal plant looks attractive in borders and mixed in among vegetables or other plants. It can also be used in mixed planters and container gardens. Calendula is a cold-hardy annual and often continues flowering, even through a layer of snow, until the ground freezes completely. Calendula is not a good companion plant in a container with other blooming annuals because it blooms better in poor soil. Fertilizing a potted calendula will give you long, tall stems with few flowers.

Recommended

C. officinalis is a vigorous, tough, upright plant that grows 12–24" tall and spreads 10–20". It bears single or double, daisy-like flowers in a wide range of yellow and orange shades. **'Bon Bon'** is a dwarf plant that grows 10–12" tall and bears flowers in shades of yellow, orange and apricot. **'Pacific Beauty'** is an heirloom cultivar with large, brightly colored flowers. It grows 18–24" tall. **'Indian Prince'** bears large, double, burnt orange flowers with mahogany centers on plants that grow 24" tall.

Problems and Pests

Calendula is usually trouble free. It continues to perform well even when afflicted with rare problems such as aphids, whiteflies, smut, powdery mildew and fungal leaf spot.

Calendula flowers are popular kitchen herbs that can be added to stews for color or to salads for flavor.

Caraway

Carum

Features: hardy biennial; feathery, light green foliage; clusters of tiny, white flowers; edible seeds **Height:** 8–24" **Spread:** 12"

In use for over 5000 years, caraway is a tasty addition to savory and sweet dishes. The seeds are used in sauerkraut, stews, rye bread, pies and coleslaw.

Starting

Caraway can be started from seed and should be planted where you want it to grow because it can bolt (go to flower) quickly when the roots are disturbed.

Growing

Caraway grows best in **full sun** but tolerates some shade. The soil should be **fertile, loose** and **well drained**. This plant is biennial and generally doesn't bloom until the second year. It is a vigorous self-seeder.

Harvesting

Cut the ripe seed heads from the plants and place them in a paper bag. Loosely tie the bag closed and hang it in a dry location. Once the seeds are dry, they can be stored in an airtight jar. The seed heads are ripe when the seedpods just begin to open.

Tips

Caraway doesn't have a very strong presence in the garden, but it can be planted in groups with more decorative plants, where its ferny foliage and pretty white flowers add a delicate, airy touch.

Recommended

C. carvi is a delicate-looking, upright biennial with ferny foliage. In the second summer after sowing, white flowers are borne in flat-topped clusters. This plant is dependably hardy to zone 5 and can be overwintered in colder gardens with a good mulch in fall.

Problems and Pests

Caraway rarely suffers from any pests or problems.

C. carvi

Plant caraway near cabbage to attract predatory insects and to help you to remember to add some caraway seed to your cabbage as it cooks. Caraway helps to reduce the cooking odor and flatulence associated with cabbage.

Carrots

Daucus

Features: biennial grown as an annual; feathery foliage; edible root
Height: 8–18" Spread: 2–4"

There is nothing quite as satisfying as pulling a perfect, crisp, sweet carrot out of the garden and crunching away as you go about your work. With all of the different sizes and colors available, carrots are a fun crop to grow with children.

Starting

Carrots can be sown directly into the garden once the last frost date has passed and the soil has warmed up. The seeds are very tiny and can be difficult to plant evenly. Mix the tiny seeds with sand before you sow them to spread the seeds more evenly and to reduce the need for thinning. You can also purchase seeds that have been coated with clay to make them easier to handle. Cover only very lightly with sand or compost when planting because the seeds can't sprout through too much soil. Keep the seedbed moist to encourage even germination.

Growing

Carrots grow best in **full sun**. The soil should be **average to fertile, well drained** and **deeply prepared**. Because you are growing carrots for the root, you need to be sure that the soil is loose and free of rocks to a depth of 8–12". This gives carrots plenty of space to develop and makes them easier to pull up when they are ready for eating. If your soil is very rocky or shallow, you may wish to plant carrots in raised beds to provide deeper, looser soil, or grow shorter varieties of carrots.

Spacing carrots is a gradual process. As the carrots develop, pull a few of the more crowded ones out, leaving

room for the others to fill in. This thinning process will give you an indication of how well they are developing and when they will be ready to harvest. The root can be eaten at all stages of development.

Harvesting

Never judge carrots by their greens. Big, bushy tops are no indication that carrots are ready for picking. As the roots develop, you will often see the top of the carrots at or just above soil level, which is a better indication of their development. You can pull carrots up by getting a good grip on the greens in loose enough soil, but you may need a small garden hand fork or large fork to dig them up in a heavier soil.

Tips

Carrots make an excellent ornamental grouping or edging plant. The feathery foliage provides an attractive background for flowers and plants with ornamental foliage.

Recommended

D. carota* var. *sativus forms a bushy mound of feathery foliage. It matures in 50 to 75 days. The edible roots may be orange, red, yellow, white or purple. They come in a variety of shapes, from long and slender to short and round. The type you choose will depend on the flavor you like, how long you need to store them and how suitable your soil is. **'Little Finger'** produces baby carrots. **'Napoli'** is an early maturing, sweet carrot. **'Purple Haze'** has purple-skinned, orange-fleshed carrots. **'Rainbow'** produces roots in white, yellow and several shades of orange. **'Thumbelina'** develops small, 2" round roots, which grow very well in poor soil or containers. **'Danvers Half-Long'** develops tasty, deep orange roots that grow well in clay soil and are suitable for overwintering. **'Scarlet Nantes'**

('Nantes') is a sweet heirloom carrot. **'Merida'** has a been bred to overwinter near the coast.

Problems and Pests

Carrot rust flies and root maggots can sometimes be troublesome but can be controlled with Reemay row covers.

To keep carrots for a long time, it is best to store them in a cold, frost-free place in containers of moistened sand.

Cauliflower

Brassica

Features: annual; white, purple, green, yellow or orange, edible flowerheads
Height: 18–24" **Spread:** 18"

If you think of pure white heads when you think of cauliflower, you may be surprised by all of the different colors available through seed vendors. Orange, purple, green and the classic white are all readily available and grow well in the Washington and Oregon.

Starting

Cauliflower can be sown directly into the garden around the last frost date, or you can start it indoors about four weeks before you plan to set it outdoors.

Growing

Cauliflower grows best in **full sun**. The soil should be **fertile, moist** and **well drained**. This plant doesn't like persistently hot weather (more than two weeks with temperatures over 80° F), which can be a problem east of the Cascades. Cauliflower must have a rich soil that stays evenly moist, or heads may form poorly, if at all. Mix plenty of compost into the soil, and mulch with compost to help keep the soil moist.

Harvesting

Unlike broccoli, cauliflower does not develop secondary heads once the main one is cut. Cut the head cleanly from the plant when it is mature. You can then compost the plant.

Tips

White cauliflower may turn yellow or greenish unless some of the leaves are tied over the head to shade it from the sun. Tie some of the outer leaves together over the head with elastic or string when you first notice it forming.

Unlike white cauliflower, the purple-, green-, yellow- or orange-headed varieties need no shading while they develop. Purple cauliflower generally turns green when cooked.

Recommended

B. oleracea* var. *botrytis is leafy and upright with dense, edible flower clusters in the center of the plant. Most selections take 70 to 85 days to mature, though some mature in as few as 45 days. Popular white-headed cultivars include **'Early Dawn,' 'Snow King,'** which is heat-tolerant, and **'Amazing.' 'Graffiti'** and **'Violet Queen'** are purple-headed cultivars. **'Panther'** and **'Veronica'** (another heat-tolerant selection) are green-headed cultivars, with Veronica's florets forming pointed, tapering peaks. **'Cheddar'** is an orange-headed cultivar.

Problems and Pests

Cutworms, leaf miners, caterpillars, root maggots, aphids, cabbage white butterfly larvae, white rust, downy mildew and powdery mildew can occur.

Celery & Celeriac

Apium

Features: biennial grown as an annual; bushy habit; edible stems; bright green leaves **Height:** 18" **Spread:** 10"

If you think celery isn't worth growing in the home garden, think again. Crisp, flavorful ribs of garden-fresh celery are far superior to what is available in the grocery store, but this vegetable is a challenge for home gardeners as it does have specific temperature requirements.

Starting

Start celery seed indoors at least eight weeks and up to 12 weeks before you plan to transplant it outdoors. When raising plants from seed, you must keep the temperatures above 55° F but below 70° F. Be patient; seed can take up to three weeks to germinate. Be sure to keep the planting medium moist, but not soggy. Do not plant out until the last frost date has passed and the soil has warmed. Mulch plants to conserve moisture.

Growing

Celery and celeriac grow best in **full sun** but enjoy light or afternoon shade in hot weather. The soil should be **fertile, humus rich, moist** and **well drained**. Allowing the soil to dry out too much will give you a poor quality, bad-tasting vegetable. Celery is one of the heaviest feeding

Cold nights can cause celery to flower, leaving the stalks inedible, or at least unpalatable.

plants in the garden. Plan on giving this vegetable extra fertilizer.

In late summer, you can mound soil around the celery stalks or wrap them in newspaper to shade them from the sun and to encourage the development of the pale green or blanched stems we are familiar with. Unblanched stalks have a stronger flavor that some people prefer. Celeriac needs no blanching.

Harvesting

Celery stalks can be harvested one or two at a time from each plant for most of summer. If you are blanching the stems before picking all or some of them, they will be ready for harvesting two or three weeks after the stalks are covered. A light touch of frost can sweeten the flavor of the celery.

Celeriac should be harvested before the first fall frost. Pull plants up, remove the leaves and store the knobby roots the same way you would beets or carrots, in a cold, frost-free location in a container of moistened sand.

Tips

Celery and celeriac have light green leaves that create a very bushy backdrop for flowering plants with less attractive, spindly growth.

Recommended

A. graveolens* var. *dulce (celery) is a bushy, upright plant with attractive, light to bright green foliage. It matures in 100 to 120 days. **'Utah 52-70'** is a disease resistant cultivar, and **'Giant Red'** is an heirloom variety with a blush of red on the stems.

A. graveolens* var. *rapaceum (celeriac) forms a bushy, bright green plant that develops a thick, knobby, bulbous root. It matures in 100 to 120 days. **'Brilliant'** and **'Giant Prague'** are popular cultivars.

Problems and Pests

Problems with fungal blight, mosaic virus, *fusarium* yellows, bacterial and fungal rot, leaf spot and caterpillars can occur.

Blanching is a gardening technique usually used on bitter-tasting vegetables. Partially or completely depriving the plant of light sweetens the flavor. Although many devices have been developed to make blanching easier, a good mound of soil or mulch will do the job just fine.

Chamomile

Matricaria

Features: bushy annual; fragrant, feathery foliage; daisy-like flowers
Height: 12–24" Spread: 6–8"

Chamomile's pretty, flavorful flowers can be used to make a perfect after-dinner tea. This delicate, airy plant is useful for filling in garden spaces wherever it is planted. The flowers can also be dried and stored so that you can enjoy chamomile tea through the winter months.

Starting

Seeds can be started about four weeks early indoors or sown directly in the garden once the last frost date has passed.

Growing

Chamomile grows best in **full sun**. The soil should be of **average fertility, sandy** and **well drained**. Chamomile self-seeds freely if you don't pick all the flowers.

Harvesting

As the flowers mature, the petals fall off and the centers swell as the seeds start to develop. At this point, the flowers can be picked and used fresh or dried for tea.

Tips

Chamomile is an attractive plant to use along the edge of a pathway or to edge beds where the fragrance will be released when the foliage is brushed up against, bruised or crushed. The flowers attract beneficial insects, so you may also want to plant a few here and there among your other plants.

Recommended

M. recutita (German chamomile) is an upright annual with soft, finely divided, fern-like foliage and bears small, daisy-like flowers in summer. It grows 12–24" tall and spreads 6–8".

Problems and Pests

Chamomile rarely suffers from any problems.

Chamomile tea is a relaxing beverage to have before bed or after a meal to help settle the stomach.

Chard
Swiss Chard
Beta

Features: biennial grown as an annual; glossy, green, purple, red or bronze, edible foliage; green, white, yellow, orange, pink, red or purple stems and veins Height: 8–18" Spread: 12"

Chard is one of the most useful vegetables to include in your garden. The tasty leaves and stems can be harvested all summer, and the wide range of colors makes it a valuable ornamental addition to beds, borders and even container gardens.

'Bright Lights'

Starting

The corky, wrinkled seed of chard is actually a dry fruit that contains several tiny seeds. Plant the dry fruit directly in the garden around the last frost date. Even if you space the seeds 3–6" apart, may have to thin the plants out a bit because several plants can sprout from each fruit.

Growing

Chard grows well in **full sun** or **partial shade**. It grows best in cool weather. The soil should be **fertile, moist** and **well drained**. Mulch lightly with compost to maintain moisture and to improve the soil.

Harvesting

Chard matures quickly, and a few leaves can be plucked from each plant every week or so all summer. You can generally start picking leaves about a month after the seed sprouts and continue to do so until the plant is killed back by frost.

Tips

Chard has very decorative foliage. Although the leaves are all generally glossy green, the stems and veins are often brightly colored in shades of red, pink, white, yellow or orange. When

planted in small groups in your borders, chard adds a colorful touch. The bushy, clumping habit also makes it well suited to mixed container plantings.

Recommended

B. vulgaris* subsp. *cicla forms a clump of glossy, green, purple, red or bronze leaves that are often deeply crinkled or savoyed. Stems and veins may be pale green, white, yellow, orange, pink, red or purple. Plants grow 8–18" tall and spread about 12". Popular cultivars include

'Fordhook Giant,' a white-stemmed heirloom; **'Oriole Orange,'** with orange stems; **'Perpetual,'** a non-bolting, heat-resistant cultivar with pale green stems; **'Bright Lights,'** with a combination of red, orange, yellow or white stems; **'Rhubarb,'** with bright red stems; and **'Golden,'** a specialty heirloom.

Problems and Pests

Rare problems with downy mildew, powdery mildew, leaf miners, aphids, caterpillars and root rot can occur.

If you find that chard fades out in your garden during the heat of summer, you can plant a second crop in mid-summer for fresh leaves in late summer and fall. In mild winters west of the Cascades, your Swiss chard will overwinter if planted in August.

Chives

Allium

Features: foliage; habit; mauve, pink or white flowers **Height:** 8–24" **Spread:** 12" or more

The delicate onion flavor of chives is best enjoyed fresh. Mix chives into dips, or sprinkle them on salads and baked potatoes. The blooms are striking in the garden and in salads and herbal vinegars.

A. schoenoprasum

Starting

Chives can be started indoors four to six weeks before the last frost date or planted directly in the garden. They also can be purchased as plants.

Growing

Chives grow best in **full sun**. The soil should be **fertile, moist** and **well drained**, but chives adapt to most soil conditions. Plants self-seed freely in good growing conditions. The youngest leaves are the most tender and flavorful, so cut plants back to encourage new growth if flavor diminishes over summer.

Harvesting

Chives can be snipped off with scissors all spring, summer and fall, as needed. Flowers are usually used just after they open, and the individual flowers in the cluster are broken apart for use.

Tips

Chives are decorative enough to be included in a mixed or herbaceous border and can be left to naturalize. In a herb garden, chives should be given plenty of space to allow for self-seeding.

Recommended

A. schoenoprasum (chives) forms a clump of bright green, cylindrical leaves. Clusters of pinkish purple flowers are produced in early and mid-summer. Varieties with white or pink flowers are also available. (Zones 3–8)

A. tuberosum (garlic chives) forms a clump of long, narrow, flat, dark green leaves. Clusters of white flowers are borne all summer. The young leaves have a distinctive garlic flavor. They are even more prone to self-seeding than chives. (Zones 3–8)

Problems and Pests

Chives rarely have any problems.

Chives are said to increase appetite and encourage good digestion.

Coriander & Cilantro

Coriandrum

Features: habit; foliage; flowers; seeds Height: 18–24" Spread: 8–18"

Coriander is a multi-purpose herb. The leaves, called cilantro, are used in salads, salsas and soups. The seeds, called coriander, are used in cakes, pies, chutneys and marmalades. The flavor of each is quite distinct.

Starting

Coriander can be started from seed four to six weeks before the last frost date or sown directly in the garden. Started plants can also be purchased from nurseries, garden centers or herb specialists. Several small sowings two weeks apart will ensure a steady supply of leaves.

Growing

Coriander prefers **full sun** but benefits from afternoon shade during the heat of summer in hotter parts of Washington and Oregon. The soil should be **fertile, light** and **well drained**. This plant dislikes humid conditions and does best during a dry summer.

Harvesting

Harvest leaves as needed throughout summer. Seeds can be harvested as they ripen in fall. Spread out a large sheet and shake the seed heads over it to collect the seeds.

Tips

Coriander has pungent leaves and is best planted where people who don't appreciate the scent won't have to brush past it. The plant is a delight to behold when in flower. Add a plant here and there throughout your borders, both for the visual appeal and to attract beneficial insects.

Recommended

C. sativum is an annual herb that forms a clump of lacy basal foliage, above which large, loose clusters of tiny, white flowers are produced. The seeds ripen in late summer and fall.

C. sativum

Problems and Pests

This plant rarely suffers from any problems.

Coriander is one of the best plants for attracting predatory insects to your garden.

Corn

Zea

Features: annual grass; broad, strap-shaped leaves; tassel-like flowers
Height: 4–8' **Spread:** 12–24"

Corn originated in South America. Widely grown by Native peoples in both North and South America, corn is one of the 'Three Sisters' of native gardens. Corn, beans and squash were grown together as companions. The beans fixed nitrogen in the soil and could climb up the corn for support. The large leaves of squash shaded the soil and kept weeds to a minimum.

Starting

Start seed directly in the garden once the last frost date has passed and the soil has warmed; seed will rot in too cool a soil, and soil temperature must be above 60° F before you direct seed. You can take the temperature of your soil with a compost thermometer. Depending on the variety you choose, corn can take from 75 to 110 days to mature. If your growing season is short or the soil is slow to warm, such as in gardens near the coast, you may prefer to start your corn four to six weeks early in peat pots and transplant it to the garden several weeks after the last frost date when the soil is warm.

Growing

Corn grows best in **full sun**. The soil should be **very fertile, moist** and **well drained**. Corn is a heavy feeder, so remember to fertilize as the plant develops. You can mound

There are ornamental varieties of corn available; some, such as 'Fiesta' and 'Seneca Indian,' have colorful kernels on the cobs; others, such as 'Harlequin' and 'Variegata,' have foliage striped in red, green and white or creamy white, respectively.

more soil around its base; the stem will develop roots in this soil, and the plant will be stronger and less likely to blow over in a strong wind.

Harvesting

Corn is ready to pick when the silks start to turn brown and the kernels are plump. If you peel the husk back just slightly, you will be able to see if the kernels are plump. Heirloom varieties should be picked and used as quickly as possible because they begin to turn starchy as soon as they are picked. Newer selections have had their genes modified to increase their sweetness and ability to stay sweet after picking. These can be stored for a while with no loss of sweetness.

Tips

Corn is wind pollinated, so plants need to be fairly close together for pollination to occur. Planting in groups improves your pollination

rates when you are growing only a few plants. Corn is an architectural grass that looks similar to some of the upright ornamental grasses, such as miscanthus. Plant it in groups of five or nine in your beds and borders. In the vegetable garden, blocks of corn should be at least 4' by 4' for best pollination.

Recommended

Z. mays is a sturdy, upright grass with bright green leaves with undulating edges. Plants grow 4–8' tall and spread 12–24". **Var. *rugosa*** (sweet corn) matures in 65 to 80 days and falls into several sweetness categories. These gauge both how sweet the corn is and how quickly the sugar turns to starch once the corn is picked. Some of the sweeter corns are less tolerant of cold soils. Kernels can be white or yellow, or cobs may have a combination of both colors. **'Earlivee'** is a cool season variety that matures in just 60 days but does not have the sweetness of a longer maturing corn like the super sweet **'Kandy Korn.' 'Early Sunglow'** does especially well in cool summer areas west of the Cascades. **'Peaches and Cream'** is another corn suited to cooler summers; it has colorful kernels. **Var. *praecox*** (popcorn) matures in 100 to 110 days and has shorter cobs with yellow, white or red, hard kernels.

Problems and Pests

Corn earworms, aphids, caterpillars, downy mildew, rust, smut and fungal leaf spots can be a problem for corn.

Avoid growing both popcorn and sweet corn, or keep them well separated, to avoid cross-pollination, which can make both inedible.

Cucumbers

Cucumis

Features: trailing or climbing, annual vine; decorative leaves; yellow flowers; edible fruit Height: 12", when trailing Spread: 4–8', when trailing

Whether you have a passion for pickling or think a salad just isn't a salad without the cool crispness of cucumber, there is a wide range of cucumbers that grow well in Washington and Oregon gardens. And there's no doubt about it: homegrown cucumbers have a crispness that is far superior to anything you can buy at your local grocer.

C. sativus (above & below)

Starting

Cucumbers can be started indoors about four weeks before the last frost date or sown directly into the garden once the last frost date has passed and the soil has warmed up. If you are starting your plants indoors, plant the seeds in peat pots so the roots will not be damaged or disturbed when you transplant them outdoors.

Wondering about the round, yellow cucumber in the photo on p. 118? 'Lemon' is a heritage variety with a crisp texture and mild flavor; it's great in salads. The vines are trailing, and because the fruit is not too heavy, they are ideal for growing up a low trellis.

Growing

Cucumbers grow well in **full sun** or **light shade**. The soil should be **fertile, moist** and **well drained**. Consistent moisture is most important when plants are germinating and once fruit is being produced. If you are growing your cucumbers up a trellis or other support, you will probably need to tie the vines in place. Use strips of old nylons or other soft ties to avoid damaging the plant.

Harvesting

When to harvest will depend on the kind of cucumbers you are growing. Pickling cucumbers are picked while they are young and small. Slicing cucumbers can be picked when mature or when small if you want to use them for pickling. Long and slender Oriental cucumbers are picked when mature and tend to be sweeter, developing no bitter flavor with age.

Tips

Cucumbers are versatile trailing plants. They can be left to wind their way through the other plants in your garden or grown up trellises or other supports. The mound-forming varieties make attractive additions to a container garden.

Recommended

C. sativus is a trailing, annual vine with coarse-textured leaves and bristly stems. It matures in 45 to 60 days. Popular slicing and salad cucumbers include the long, slender **'English Telegraph'** and **'Sweet Success'**; All America Selections winner **'Fanfare,'** a prolific bush selection; and the high-yielding, space-saving **'Spacemaster.'** Popular cultivars of pickling cucumbers include the disease-tolerant, semi-bush **'Cross Country'** and the prolific **'Pickalot.'** In warm summer areas east of the Cascades, try the **'Armenian'** cucumber relative that grows 24–36" long and will produce at higher temperatures.

Problems and Pests

Problems with powdery mildew, downy mildew, mosaic virus, white flies, aphids, cucumber beetles, bacterial wilt, leaf spot, scab and ring spot can occur.

Cucumbers keep producing fruit as long as you don't let the fruit stay on the vines for too long. Pick cucumbers as soon as they are a good size for eating. The more you pick, the more the plants will produce.

Dill

Anethum

Features: feathery, edible foliage; yellow, summer flowers; edible seeds
Height: 2–5' **Spread:** 12" or more

Dill is majestic and beautiful in any garden setting. Its leaves and seeds are probably best known for their use as pickling herbs, though they have a wide variety of other culinary uses.

Starting

Dill can be sown directly into the garden around the last frost date. Make several small sowings every couple of weeks to ensure a steady supply of leaves.

Growing

Dill grows best in **full sun** in a **sheltered** location out of strong winds. The soil should be of **poor to average fertility, moist** and **well drained**. Don't grow dill near fennel because the two plants will cross-pollinate, and the seeds of both plants will lose their distinct flavors.

Harvesting

Pick the leaves as needed throughout summer and dry them for use in winter. Harvest the seeds by shaking the seed heads over a sheet once they ripen in late summer or fall.

Tips

With its feathery leaves, dill is an attractive addition to a mixed bed or border. It can be included in a vegetable garden but does well in any sunny location. Dill also attracts butterflies and beneficial insects to the garden.

Recommended

A. graveolens is an annual herb that forms a clump of feathery foliage. Clusters of yellow flowers are borne at the tops of sturdy stems.

Problems and Pests

Dill rarely suffers from any problems.

Fennel

Foeniculum

Features: short-lived perennial or annual; attractive, fragrant foliage; yellow, late-summer flowers; edible seeds **Height:** 2–6' **Spread:** 12–24"

All parts of fennel are edible and have a distinctive, licorice-like fragrance and flavor. The seeds are commonly used to make a tea that is good for settling the stomach after a large meal. Florence fennel produces a large, swollen base that is eaten raw in salads, cooked in soups and stews or roasted like other root vegetables.

Starting

Seeds can be started directly in the garden around the last frost date or about four weeks earlier indoors.

Growing

Fennel grows best in **full sun**. The soil should be **average to fertile, moist** and **well drained**. Avoid planting fennel near dill or coriander because cross-pollination reduces seed production and makes the seed flavor of each less distinct. Fennel easily self-sows and can be invasive here.

Harvesting

Harvest fennel leaves as needed for fresh use. The seeds can be harvested when ripe, in late summer or fall. Shake the seed heads over a sheet to collect the seeds. Let them dry before storing them. Florence fennel can be harvested as soon as the bulbous base becomes swollen. Pull plants up as needed, and harvest any left in the ground before the first fall frost.

Tips

Fennel is an attractive addition to a mixed bed or border. The flowers attract pollinators and predatory insects to the garden.

Recommended

F. vulgare is a short-lived perennial that forms clumps of loose, feathery foliage. Clusters of small, yellow flowers are borne in late summer, followed by seeds that ripen in fall. **Var. *azoricum*** (Florence fennel, anise) is a biennial that forms a large, edible bulb at the plant base. (Zones 4–9)

Problems and Pests

Fennel rarely suffers from any problems.

Bronze fennel is attractive as well as delicious. The feathery bronze foliage adds a distinctive touch to flower beds and borders.

Fiddlehead Ferns

Ostrich Fern

Matteuccia

Features: perennial fern; attractive foliage; edible, tender shoots **Height:** 3–5' **Spread:** 12–36"

These popular, classic ferns are revered for their delicious, emerging spring fronds and their stately, vase-shaped habit.

The tightly coiled, new spring fronds taste delicious lightly steamed when served with butter. Remove the bitter, reddish brown, papery coating before steaming.

M. struthiopteris

Starting

Crowns can be purchased and planted out in spring. Plants are sometimes also available in summer. They are often sold as ornamental plants in the perennial department of your local garden center or nursery.

Growing

Fiddlehead ferns prefer **light or partial shade** and tolerate full shade, or full sun if the soil stays moist. The soil should be **average to fertile, humus rich, neutral to acidic** and **moist.** Leaves may scorch if the soil is not moist enough. These fronds are aggressive spreaders that reproduce by spores; give them a large area to spread so you'll have plenty of fronds to harvest.

Harvesting

Let plants become established for a couple of years before you begin harvesting. Pick new fronds in spring as they are just beginning to uncurl, and be sure to pick no more than half the fronds from each plant.

Tips

These ferns appreciate a moist woodland garden and are often found growing wild alongside woodland streams and creeks. Useful in shaded borders, these plants are quick to spread.

Recommended

M. struthiopteris (*M. pensylvanica*) forms a circular cluster of slightly arching, feathery fronds. Stiff, brown, fertile fronds, covered in reproductive spores, stick up in the center of the cluster in late summer and persist through winter. They are popular choices for dried arrangements.

Problems and Pests

These ferns rarely suffer from any problems.

Flax

Linum

Features: upright annual; blue flowers; edible seeds **Height:** 12–36" **Spread:** 8–18"

Flaxseed has long been added to baked muffins, breads and cereals. It is now being hailed for its fantastic health-improving potential and is an excellent source of omega-3 fatty acids.

L. usitatissimum (above & below)

Starting

Start seed directly in the garden around the last frost date.

Growing

Flax grows best in **full sun**. The soil should be of **average fertility, light, humus rich** and **well drained**.

Harvesting

When they are ripe, harvest the seeds by rubbing the seed heads between your hands over a sheet or bucket. Dry the seeds completely before storing them in a cool, dry place.

Tips

Flax is a beautiful plant that many gardeners grow for its ornamental appeal alone. Each flower lasts only one day and is replaced by another each day once blooming begins.

Recommended

L. usitatissimum forms clumps with leafy, upright stems that wave at the slightest breeze. The blue, summer flowers are followed by chestnut brown, pale brown or golden yellow seeds in late summer or fall.

Problems and Pests

Problems with rot, rust, wilt, slugs, snails and aphids can occur.

This species is not used exclusively as a food source; the stems of some cultivars are processed to produce linen. Linseed oil also comes from the seeds of flax.

Garlic

Allium

Features: perennial bulb; narrow, strap-like leaves; white, summer flowers
Height: 6–24" **Spread:** 8"

Garlic is not the most ornamental plant, but the light green leaves make a good groundcover and repel several common garden pests.

Starting

Garlic is generally grown from sets (cloves) that can be started in fall or spring. In the wet winter areas near the coast, always plant garlic cloves in October to prevent rotting. Fall-planted garlic produces the largest bulbs at harvest time, but don't expect to see much development from this bulb until the following spring. All the fall growth will be underground, and gives the plant a head start in spring.

Growing

Garlic grows best in **full sun**. The soil should be **fertile, moist** and **well drained**. Near the coast, garlic will need to be planted in a raised bed for improved drainage.

Harvesting

This plant can be dug up in fall once the leaves have yellowed and died back. Softneck garlic can be stored for a longer time than hardneck varieties, but it is not as hardy.

Tips

Although the flowers are quite attractive and often intriguing, they should be removed so the plant devotes all its energy to bulb production rather than seed production. Garlic takes up very little space and can be tucked into any spare spot in your garden.

Recommended

A. sativum* var. *ophioscordon (hardneck garlic) has a stiff central stem around which the cloves develop.

A. sativum* var. *sativum (softneck garlic) develops more cloves but has no stiff stem. The soft leaves are often used to braid garlic bulbs together for storage.

Problems and Pests

A few rot problems can occur, but this plant is generally trouble free.

Hardy Kiwi

Actinidia

Features: early-summer flowers, edible fruit, deciduous, woody, climbing twining habit **Height:** 15–30' **Spread:** 15–30'

Hardy kiwi is a handsome vine. Its lush green leaves and adaptability make it very useful, especially on difficult sites. With a vigorous growth habit, kiwi can swallow a small arbor with exuberance in a single season. Remember that kiwis need a male vine for pollination, if you want fruit.

Starting

Not recommended; purchase plants in containers from a nursery.

Growing

Hardy kiwi vines grow best in **full sun** and tolerate partial shade. The soil should be **fertile** and **well drained**. They require shelter from strong winds.

Prune in late winter. Plants can be trimmed to fit the area they've been given, or, if greater fruit production is desired, cut back side shoots to two or three buds from the main stems.

Harvesting

Harvest fruit when they are still firm in September, but let them finish ripening indoors rather than allowing them to ripen on the vine. The fruit is ripe when it is sweet to taste. Eat the fruit skin and all.

Tips

Kiwi vines need a sturdy structure to twine around. Pergolas, arbors and sufficiently large and sturdy fences provide good support. Given a trellis against a wall, tree or some other upright structure, hardy kiwis will twine upward all summer. They can also be grown in containers. They can grow uncontrollably; prune them back if they get out of hand.

Recommended

A. arguta (hardy kiwi, bower actinidia) grows 20–30' tall but can be trained to grow lower with judicious use of pruning shears. It has dark green, heart-shaped leaves, white flowers and smooth-skinned, greenish yellow, edible fruit.

A. kolomikta (variegated kiwi vine, kolomikta actinidia) grows 15–20' tall. It has green leaves strongly variegated with white and pink, white flowers and smooth-skinned, greenish yellow, edible fruit.

Problems and Pests

Occasional afflictions with fungal diseases are not a serious concern.

Male and female kiwi plants must be grown together to produce fruit. Containers are often sold with plants of both sexes in the same pot.

These species are good substitutes for the commercially available brown, hairy-skinned kiwi (A. chinensis; A. deliciosa). *The fruits of the species described here are hairless and high in Vitamin C, potassium and fiber.*

Horseradish

Armoracia

Features: clump-forming perennial; pungent, edible root; glossy, green, creased leaves **Height:** 24–36" **Spread:** 18" or more

Horseradish is a plant that is too often relegated to a neglected back corner of the garden. Despite its rampant ability to spread, it has beautiful foliage that makes it a prime candidate for perennial and mixed borders.

Horseradish sauce is a popular garnish for roasted meats, roast beef in particular.

Starting

Plants can be purchased, or if you know someone with a horseradish plant, they can give you a division or a piece of root. As with dandelions, a new plant will generally grow from even a small piece of root.

Growing

Horseradish grows well in **full sun**. The soil should be **fertile, moist** and **well drained**, but the plant adapts to most conditions. This plant does especially well in areas with cool summers.

Harvesting

When the foliage dies back in fall, you can dig up some of the roots to use fresh or in preserves, relishes and pickles. The roots have the strongest flavor in fall, winter and early spring, but they can be harvested at any time once the plant is well established.

Tips

Horseradish is a vigorous plant that spreads to form a sizeable clump. It is fairly adaptable and can be used in a somewhat neglected area, but it deserves a better spot because of its glossy, creased foliage. Just be sure you contain the roots with frequent shovel pruning or this plant will take over your yard.

Recommended

A. rusticana forms a large, spreading clump of large, puckered, dark green leaves. Plants grow up to 36" tall and spread 18" or more. (Zones 3–8)

Problems and Pests

Generally problem free, horseradish can occasionally suffer from powdery mildew, downy mildew, fungal leaf spot or root rot.

A. rusticana

Kale, Collards & Mustard Greens

Brassica

Features: tender biennial; bronze, purple, blue-green or glossy green, sometimes deeply wrinkled, decorative, edible leaves **Height:** 18–24" **Spread:** 18–24"

These nutrient-packed, leafy cabbage and broccoli siblings are some of the most decorative members of this family. Collards are not grown as often in Washington and Oregon as they are in the South, but they grow quite happily here and make a welcome change from cabbage and kale. Mustards are most often eaten when quite young and lend a spicy flavor to stir-fries and salads.

Starting

All three of these plants can be sown directly into the garden. They are all quite cold hardy and can be planted pretty much as soon as the soil can be worked in spring. A light frost won't harm them, though you may wish to cover the young plants if nighttime temperatures are expected to get to 15° F or lower. You may wish to make several successive plantings of mustard because the leaves have the best flavor when they are young and tender. Kale can also be planted in late July for a winter harvest in the mild winter areas near the coast and Puget Sound.

Growing

These plants grow best in **full sun**. The soil should be **fertile, moist** and **well drained**. Mustard in particular should not be allowed to dry out, or the leaves may develop a bitter flavor.

Harvesting

Once plants are established, you can start harvesting leaves as needed. Pick a few of the outer leaves from each plant.

Tips

Because the leaves are what you will be eating, you don't have to wait very long after planting to start harvesting. These plants make a striking addition to beds and borders, where the foliage creates a good complement and backdrop for plants with brightly colored flowers.

Recommended

B. oleracea* subsp. *acephala (kale, collards) and ***B. juncea* subsp. *rugosa*** (mustard) form large clumps of ruffled, creased or wrinkled leaves in shades of green, blue-green, bronze and purple. **'Lacinato'** and **'Red Russian'** are popular selections of kale. **'Green Glaze'** is a popular selection of collards. **'Mizouna'** and **'Savanna'** are popular mustard selections.

Problems and Pests

Problems with cutworms, leaf miners, caterpillars, root maggots, cabbage white butterfly larvae, white rust, downy mildew and powdery mildew can occur. Cover seedlings with Reemay floating row covers to keep out caterpillars and root maggots and protect new seedlings from slugs.

Kohlrabi

Brassica

Features: biennial grown as an annual; pale silvery or gray-green foliage; edible, bulbous stem base **Height:** 12" **Spread:** 12"

The tender, swollen stem base of kohlrabi has a very strange appearance. The leaf bases become stretched as the bulb forms, and new leaves continue to sprout from the top of the rounded bulb.

Encourage good growth by keeping the soil moist, and harvest quickly once bulbs form.

Starting

Seed can be sown directly in the garden around the last frost date. This plant matures quite quickly, so make several small sowings one or two weeks apart to have tender, young kohlrabi for most of summer. West of the Cascades, seed kohlrabi in mid-July for harvest up until December.

Growing

Kohlrabi grows best in **full sun**. The soil should be **fertile, moist** and **well drained**, though plants adapt to most moist soils.

Harvesting

Keep a close eye on your kohlrabi because the bulb can become tough and woody quickly if left too long before harvesting. The bulb is generally well rounded and 2–4" in diameter when ready for harvesting. Pull up the entire plant and cut just below the bulb. Then cut the leaves and stems off and compost them or use them to mulch the bed.

Tips

Low and bushy with white or purple bulbs, kohlrabi makes an interesting edging plant for beds and borders and can be included in container gardens, particularly those you like to change the plantings in regularly.

Recommended

B. oleracea* subsp. *gongylodes forms a low, bushy clump of blue-green foliage. As the plant matures, the stem just above ground level swells and becomes rounded. This is the edible part. **'Kongo'** is a sweet variety with good keeping quality, and **'Koliri'** has beautiful purple skin and is one of the best for late-season harvests.

Problems and Pests

Problems with cutworms, leaf miners, caterpillars, root maggots, cabbage white butterfly larvae, white rust, downy mildew and powdery mildew can occur.

Leeks

Allium

Features: biennial often grown as an annual; narrow, upright growth with long, arching, strap-like, dark blue-green leaves Height: 18–24" Spread: 4–8"

Leeks can rival most ornamental grasses for garden presence. The plants are strongly upright with stunning dark blue-green leaves that arch from the main stem. Planted in a small group, they are a welcome addition to the border.

Starting

You can start leeks indoors 8 to 10 weeks before transplanting outside, but such a long time indoors can encourage lanky growth if you don't have supplemental lighting or a very bright window. It might be better to purchase plants when you are ready to plant them outdoors, or sow directly in the soil.

Growing

Leeks are a cool-season crop and do best west of the Cascades. They grow best in **full sun**. The soil should be **fertile** and **well drained**. Mix in compost or add a layer of compost mulch once you have planted. Mound mulch, soil or straw up around the base of the plants as summer progresses to encourage tender, white growth low on the plant.

Harvesting

Leeks can be harvested when mature in early fall, but you can just harvest these hardy plants as you need them. When the ground begins to freeze, pull up any plants you want for winter use. They keep for several weeks in the refrigerator if you cut the roots short and wrap the leeks in plastic. Freeze for longer storage; double bag them so the onion-like flavor doesn't spread. Near the coast, leeks can be harvested all winter. You can also leave some plants to flower and go to seed. These perennial plants will return year after year, and new seedlings will replace fading plants.

Tips

Leeks are very ornamental. They have bright blue-green leaves that cascade from the central stem. The second year after planting, they develop large, globe-shaped clusters of flowers atop 36" stems. Plant them in groups in your beds and borders.

Recommended

A. ampeloprasum* subsp. *porrum is an upright perennial that flowers the second year after planting. **'Tadorna'** is a good variety for winter gardens as it matures in October and November and can be harvested up until a hard freeze.

Problems and Pests

Rot, mildew, smut, rust, leaf spot, onion flies and thrips can occur.

Lettuce & Mesclun

Lactuca

Features: annual; ruffled leaves in shades of green, red and bronze, sometimes spotted, speckled or streaked and bicolored **Height:** 6–18" **Spread:** 6–18"

With so many types of lettuce available, you never have to eat the same salad twice. It comes in a multitude of colors, textures and flavors, and it deserves a spot in every garden. Mesclun is a blend of different varieties, whether a spicy or a mild blend, and makes a welcome addition to salads and stir-fries while providing a good groundcover in the garden.

Starting

Lettuce and mesclun can be started directly in the garden a few weeks before the last frost date. You may wish to be precise when planting head lettuces so you don't have to thin the plants out as much later, but the seeds of leafy lettuces and mesclun can be scattered across a prepared area and do not have to be planted in rows. If you make several smaller plantings spaced a week or two apart, you won't end up with more plants than you can use maturing at once. For an earlier crop, start a few plants indoors about four weeks before you plan to plant them outdoors.

Growing

Lettuce and mesclun grow well in **full sun, light shade** or **partial shade** in a **sheltered** location. The soil should be **fertile, moist** and **well drained**. Add plenty of compost to improve the soil, and be sure to keep your lettuce moist. Lettuce is very prone to drying out in hot and windy situations, so east of the Cascades, plant as early as possible so it can grow while the summer is still cool. Plants under too much heat stress can quickly bolt or flower and go to seed, which causes the greens to taste bitter.

Harvesting

Head-forming lettuce can be harvested once the head is plump. If the weather turns very hot, you may wish to cut heads even earlier because the leaves develop a bitter flavor once plants go to flower. Rinsing bitter lettuce in warm water can remove some of the strong flavor. Loose-leaf lettuce and mesclun can be harvested by pulling a few leaves off as needed or by cutting an entire plant 2–4" above ground level. Most will continue to produce new leaves even if cut this way.

Tips

Lettuce and mesclun make interesting additions to container plantings, either alone or combined with other plants. In beds and borders, mesclun makes a decorative edging plant. All lettuces and mesclun are fairly low growing and should be planted near the front of a border so they will be easier to get to for casual picking.

Recommended

L. sativa forms a clump of ruffle-edged leaves and comes in many forms. Loose-leaf lettuce forms a loose rosette of leaves rather than a central head. Butterhead or Boston lettuce forms a loose head and has a very mild flavor. Crisphead or iceberg lettuce forms a very tight head of leaves. Romaine or cos lettuce has a more upright habit, and the heads are fairly loose but cylindrical in shape.

Mesclun mixes can be a combination of different lettuces, usually looseleaf types, to be eaten while very young and tender. Mixes often also include other species of plants, including mustards, broccoli, radicchio, endive, arugula, chicory and spinach. Most seed catalogs offer a good selection of pre-mixed mesclun, as well as separate selections to create your own mix.

Problems and Pests

Problems with root rot, leaf spot, flea beetles and mosaic virus can occur.

Marigolds

Tagetes

Features: annual; fragrant foliage; yellow, red, orange, brown, gold, cream or bicolored, edible flowers **Height:** 6–36" **Spread:** 6–24"

From the large, exotic, ruffled flowers of African marigold to the tiny flowers of the low-growing signet marigold, the warm colors and fresh scent of these plants add a festive touch to the garden.

Starting

Marigolds can be sown directly in the garden around the last frost date. Plants can also be purchased from garden centers and nurseries.

Growing

Marigolds grow best in **full sun**. The soil should be of **average fertility** and **well drained**. These plants are drought tolerant but hold up well in windy, rainy weather. Deadhead to prolong blooming and to keep plants tidy.

Harvesting

Flowers should be picked for use once they are fully open. Leaves of signet, lemon mint and Mexican mint marigolds can be picked as needed for use in teas, soups and salads.

Tips

Dot these plants in small groups throughout your beds and borders for a pretty display, as well as to take advantage of their reputed nematode-repelling properties. Marigolds also make lovely additions to sunny container plantings.

Recommended

Although ***T. erecta*** (Aztec marigold, African marigold), ***T. patula*** (French marigold) and their hybrids have edible flowers, ***T. lemmonii*** (lemon mint marigold), ***T. lucida*** (Mexican mint marigold, Mexican mint tarragon) and ***T. tenuifolia*** (signet marigold) are the most common culinary species.

Problems and Pests

Slugs and snails can ravage the foliage of all marigolds.

Melons

Cucumis

Features: trailing or climbing, annual vine; attractive foliage; yellow flowers
Height: 12" **Spread:** 5–10'

Success with melons is somewhat limited in the cool summer areas west of the Cascades because the plants prefer warmer summer weather and a longer growing season. The keys to success, even in a short-season region, are to select varieties that mature quickly and to plant in the warmest, sunniest spot you can find.

Starting

Melons can be started indoors about six weeks before you want to transplant them to the garden. They don't like to have their roots disturbed, so plant them in fairly large peat pots so they have plenty of room to grow and can be set directly into the garden once the weather warms up. Do not overwater the seedlings—they are susceptible to mildew—and do not set seedlings outdoors until the soil temperature is at least 60° F.

Growing

Melons grow best in **full sun** in a warm location. In all parts of Oregon and Washington, melons will do better in raised beds with a silver or black plastic mulch to warm the soil. The soil should be **average to fertile, humus rich, moist** and **well drained**. Fruit develops poorly with inconsistent moisture, and plants can rot in cool or soggy soil. The fruit will be sweeter and more flavorful if you cut back on watering as it is ripening.

'Fastbreak'

Melons have a well-deserved reputation for spreading, but like most vine-forming plants, they can be trained to grow up rather than out. As the fruit becomes larger, you may need to support it so the vines don't get damaged. You can create hammocks out of old nylon pantyhose to support the melons.

Harvesting

Melons should be allowed to fully mature on the vine. Muskmelons develop more netting on the rind as they ripen. They generally slip easily from the vine with gentle pressure when ripe. Honeydew melons develop a paler color when they are ripe. They must be cut from the vine when they are ready.

Tips

Melons have quite attractive foliage and can be left to wind through your ornamental beds and borders if you don't want to train them to grow up.

Recommended

C. melo **subsp.** ***reticulatus*** (muskmelon) and ***C. melo*** **subsp.** ***indorus*** (honeydew melon) are tender annual vines with attractive green leaves. Bright yellow flowers are produced in summer. Male and

female flowers are produced separately on the same vine. The melons are round and green or gold, and some, usually muskmelons, develop a corky tan or greenish netting as they ripen. Muskmelons generally develop orange or salmon-colored flesh, and honeydew melons develop pale green or yellow flesh. Most melons take 70 to 85 days to produce mature, ripe fruit. Popular cultivars for short seasons include **'Alaska,' 'Earlidew,' 'Fastbreak,' 'Gourmet'** and **'Passport.'**

Problems and Pests

Powdery mildew, *Fusarium* wilt, cucumber beetles and sap beetles can be quite serious problems. Mildew weakens the plants, and the beetles may introduce wilt, which is fatal to the plants.

Nastursiums

Tropaeolum

Features: bushy or trailing annual; attractive, edible foliage; red, orange, yellow, burgundy, pink, cream, gold, white or bicolored, edible flowers **Height:** 12–18" for dwarf varieties; up to 10' for trailing varieties **Spread:** equal to height

These fast-growing, brightly colored flowers are easy to grow, making them popular with beginners and experienced gardeners alike. Nasturtiums are good edibles for kids to try.

T. majus (above)

Starting

Direct sow seed once the danger of frost has passed. Pre-soaking the seeds overnight will help them to germinate sooner.

Growing

Nasturtiums prefer **full sun** but tolerate some shade. The soil should be of **poor to average fertility, light, moist** and **well drained**. Soil that is too rich or has too much fertilizer results in a lot of leaves and very few flowers. Let the soil drain completely between waterings.

Harvesting

Pick leaves and flowers for fresh use as needed. The seedpods can be used as a substitute for capers.

Tips

Nasturtiums are used in beds, borders, containers and hanging baskets and on sloped banks. The climbing varieties are grown up trellises and over rock walls or places that need concealing. These plants thrive in poor locations, and they make interesting additions to plantings on hard-to-mow slopes.

Recommended

T. majus has a trailing habit, but many of the cultivars have bushier, more refined habits. Cultivars offer differing flower colors or variegated foliage.

Problems and Pests

Aphids are the biggest pest and can be controlled by frequent washing with a strong jet of water.

Onions

Allium

Features: biennial or perennial; upright, tubular, green leaves; globe-like flower clusters **Height:** 18–24" **Spread:** 2–4"

Onions are one of the oldest cultivated plants, having been grown for over 5000 years. They have been cultivated for so long that their country of origin is not even known. They are well worth growing because the many interesting varieties make onions a year-round crop for home gardeners.

Starting

Onions can be started from seed indoors about six to eight weeks before you plan to plant them outdoors; they can also be sown directly in the garden once the last frost date has passed. Sets of starter onions can be purchased and planted in spring. Bunching onions are usually started from seed sown directly in the garden. They are quick to mature, so make several smaller sowings two or three weeks apart from spring to mid-summer for a regular supply.

Growing

Onions should be grown in **full sun**. The soil should be **fertile, moist** and **well drained**. Onions use plenty of water but will rot in very wet soil. They are poor at competing with other plants, so keep them well weeded. A good layer of mulch will conserve moisture and keep the weeds down. Be sure to water during periods of extended drought.

Harvesting

All onions can be harvested and used as needed throughout the season. Pull up onions that need thinning, and just pinch back the tops if you want the bulbs to mature or the plant to continue to produce leaves.

Bulb onions grown for storage are ready to be harvested when the leaves begin to yellow and flop over and the shoulders of the bulbs are just visible above the soil line. They should be pulled up and allowed to dry for a few

days before being stored in a dry, cold, frost-free spot.

Onions are perennials and can also be left in the ground over winter, though the flavor can become quite strong the second year. They will flower the second summer.

Tips

Onions have fascinating cylindrical leaves that add an interesting vertical accent to the garden. Include them in beds and borders, but if you want big bulbs, don't let them be overly crowded by other

Onions make interesting additions to container plantings, particularly close to the house, if you just want to pick a few leaves at a time. The pungent smell of onions makes them a natural insect repellent. They are good companions to other garden vegetables.

A. fistulosum (above)

plants. Smaller bulbs form on plants if you want small bulbs.

Recommended

A. cepa (bulb onion) forms a clump of cylindrical foliage and develops a large, round or flattened bulb. Bulb formation is day-length dependent. It is best to use northern onion varieties that have been developed for long days. Bulb formation begins when days are 15–16 hours long. Walla Walla sweet onions mature in 125 days when planted in spring, but they can also be fall planted and allowed to overwinter.

A. fistulosum (bunching onion, green onion, shallot) is a perennial that forms a clump of foliage. The plant quickly begins to divide and multiply from the base. Plants may develop small bulbs or no bulbs at all. Once established, these plants will provide you with green onions all spring and summer.

Problems and Pests

Problems with smut, onion maggots and rot can occur.

A. fistulosum (above)

Oregano & Marjoram

Origanum

Features: fragrant foliage; white or pink, summer flowers **Height:** 12–32" **Spread:** 8–18"

Oregano and marjoram are two of the best-known and most frequently used herbs. They are popular in stuffings, soups and stews, and no pizza is complete until it has been sprinkled with fresh or dried oregano.

Starting

These plants can be started from seed four to six weeks before you plan to plant them into the garden. You can also purchase plants, but use a reputable source so you can be sure you purchase the exact variety you want.

Growing

Oregano and marjoram grow best in **full sun**. The soil should be **average to fertile, neutral to alkaline** and **well drained**. The flowers attract pollinators and beneficial insects to the garden. Oregano may self-seed.

Harvesting

Leaves can be picked as needed for fresh use or dried for use in winter.

Tips

These bushy perennials make lovely additions to any border and can be trimmed to form low hedges. Dwarf oregano makes an attractive rock garden plant.

Recommended

O. majorana (marjoram) is upright and shrubby. It has fuzzy, light green leaves and bears white or pink flowers in summer. Where it is not hardy, it can be grown as an annual. (Zones 5–9)

O. vulgare* var. *hirtum (organo, Greek oregano) is the most flavorful culinary variety of oregano. This low, bushy plant has fuzzy, gray-green leaves and bears white flowers. Many other interesting varieties of *O. vulgare* are available, including those with golden, variegated or curly leaves. (Zones 5–9)

Problems and Pests

Problems with oregano and marjoram are rare.

Oriental Cabbage

Brassica

Features: upright, leafy annual **Height:** 18" **Spread:** 12"

A tasty ingredient in stir-fries and soups, Oriental cabbage is also an interesting addition to the garden. It has light or dark green, white-veined leaves with undulating or ruffled edges.

Starting

Start seeds indoors four to six weeks before the last frost date. Transplant young plants outdoors once the last frost date has passed and the soil has warmed up. Oriental cabbage can also be planted in July and August for fall harvest.

Growing

Oriental cabbage grows best in **full sun**. The soil should be **fertile, moist** and **well drained**. Mulch to keep the soil moist. Oriental cabbage does best in a cool summer climate and will bolt in warmer areas east of the Cascades. It does especially well in gardens near the coast.

Harvesting

Oriental cabbage comes in two basic forms: solid heads that are cut whole, and looser heads with leaves that can be removed as needed.

Tips

This cabbage can be grown in containers, where it can be combined with other edible or flowering plants. Dotted through a border, it creates a low but upright feature, adding its unique color and form.

Recommended

B. rapa* subsp. *chinensis (pac choi, bok choi) forms a loose clump of blue-green leaves with thick, fleshy, white or light green stems. It is bolt resistant for warmer gardens.

B. rapa* subsp. *pekinensis (Chinese cabbage) forms a dense head of tightly packed leaves.

Problems and Pests

Cutworms, flea beetles, leaf miners, caterpillars, root maggots, cabbage white butterfly larvae, white rust, downy mildew and powdery mildew can be problematic. Cover plants with floating row covers of Reemay to keep out these pests, and rotate the crop every year.

Many seed catalogs offer a selection of Oriental vegetables. Some of them are cultivars or species of plants we recognize, while others are unique. Try a few to expand your vegetable repertoire.

Parsley

Petroselinum

Features: biennial grown as an annual; bushy habit; attractive, edible foliage
Height: 8– 24" **Spread:** 12–24"

Parsley is far more than an attractive garnish. It is flavorful and full of vitamins, and it can brighten the flavor of just about any dish.

Starting

Parsley can be sown directly in the garden once the last frost date has passed, or four to six weeks earlier indoors. Start it in peat pots or in its permanent location because parsley resents having its roots disturbed.

Growing

Parsley grows well in **full sun** or **partial shade**. The soil should be **average to fertile**, **humus rich**, **moist** and **well drained**.

Harvesting

Pinch parsley back to encourage bushy growth, and use the sprigs you pick off for eating. This plant can also be cut back regularly if you need a larger quantity in a recipe, and if not cut back too hard, it will sprout new growth.

Tips

Containers of parsley can be kept close to the house for easy picking. The bright green leaves and compact habit make parsley a good edging plant for beds and borders.

Recommended

P. crispum forms a clump of bright green, divided leaves. This biennial is usually grown as an annual because it is the leaves that are desired, not the flowers or seeds. Cultivars may have flat or curly leaves. Flat leaves are tastier, and curly ones are more decorative. Dwarf cultivars are available.

Problems and Pests

Parsley rarely suffers from any problems.

Parsnips

Pastinaca

Features: biennial grown as an annual; sweet, edible root; feathery foliage **Height:** 12–18" **Spread:** 6–8"

Despite needing a fairly long season to mature, parsnips are well suited to our gardens because they are best eaten once they have had a few good frosts to sweeten their roots.

Starting

Sow seed directly into the garden as soon as the soil can be worked. Be sure to keep the soil moist to ensure good germination. Seed can be slow to germinate, sometimes taking up to three weeks, so mark the location where you plant it. You can continue sowing parsnip seeds up until July for harvest all winter.

Growing

Parsnips grow best in **full sun** but tolerate some light shade. The soil should be of **average fertility, moist** and **well drained**. Be sure to work the soil to a depth of at least 12" because parsnips need loose soil much like carrots. Parsnips do not like high nitrogen fertilizer or fresh manure. Mix in compost to improve the texture. Roots develop poorly in heavy soil. Mulch to suppress weed growth and to conserve moisture.

Harvesting

The roots can be pulled up any time from October through February, and in fall after the first few frosts they can be stored, like carrots, in damp sand in a cold, frost-free location. These plants can also be mulched with straw and pulled up in spring before they sprout new growth. Frost improves the flavor because some of the root starches are converted to sugar in freezing weather.

Tips

Not the most ornamental of vegetables, parsnips provide a nice dark, leafy background for lower-growing plants and produce plenty of winter vegetables for very little effort.

P. sativa

Recommended

P. sativa is an upright plant with dark green, divided leaves. It develops a long, pale creamy yellow root that looks like a carrot. **'Andover,' 'Gladiator'** and **'Hollow Crown'** are commonly available cultivars.

Problems and Pests

Canker, carrot rust fly and onion maggot can affect parsnips.

Roasting the roots brings out their sweetness. Combine parsnips with potatoes, carrots and other root vegetables, then drizzle with oil and sprinkle with herbs before roasting for an hour, or until the vegetables are tender.

Peas

Pisum

Features: climbing annual; bright green foliage; white flowers **Height:** 1–5' **Spread:** 4–8"

It's easy to love peas. They are easy to grow (as long as the weather is cool), versatile, tasty and easy to store. Whether you long for the fresh flavor of shelling peas or the satisfying crunch of snap peas, they are especially easy to grow west of the Cascades in cool summer areas.

Starting

Peas show an admirable appreciation for cool spring weather and can be planted directly outdoors as soon as the soil can be worked and has dried out a bit. A light frost or two won't do them any harm, but the seeds can rot in cold, wet soil. Pre-sprout pea seeds by wrapping them in a damp dish towel. Planting the seeds with a bit of the roots poking out will keep them from rotting in wet spring weather.

Growing

Peas grow well in **full sun**. The soil should be **average to fertile, humus rich, moist** and **well drained**. Peas can grow a wide range of heights, but all benefit from a support of some kind to grow up. They develop small tendrils that twine around twiggy branches, nets or chain-link fences. Base your support height on the expected height of the plants, and make sure it is in place before your seeds sprout because the roots are quite shallow and can be damaged easily.

Harvesting

Peas should be harvested when they are still young and tender. They can be pulled from the vine by hand, but use both hands, one to hold the plant and one to pull the pea pod, to avoid damaging the plant. The more you pick, the more peas the plants will produce.

Tips

Peas are excellent plants for growing up a low chain-link fence, and the taller varieties create a privacy screen quite quickly. The low-growing and medium-height peas make interesting additions to hanging baskets and container plantings. They can grow up the hangers or supports or can be encouraged to spill over the edges and trail down.

Recommended

P. sativum* var. *sativum are climbing plants with bright green, waxy stems and leaves and white flowers. The resulting pods are grouped into three categories: shelling peas, snow peas and snap peas. The seeds are removed from the pods of shelling peas and are the only part eaten. Snow peas are eaten as flat, seedless pods. Snap peas develop fat seeds, and the pod and seeds are eaten together. **'Oregon Pioneer'** has been developed for the maritime Northwest to be resistant to mosaic disease. **'Oregon Sugar Pod II'** is another favorite for the Northwest, and **'Sugar Spring'** and **'Super Sugar Snap'** are extra sweet and do especially well. The tall growing **'Alderman'** pea matures late to extend the harvest. **'Cascadia'** is another variety bred for the Northwest.

There are many peas to choose from. When deciding which kind to grow, think about the type of peas you will get the most use from. Then choose the mature plant height that is most suitable for the space you have.

Problems and Pests

Peas are somewhat prone to powdery mildew, so choose mildew-resistant varieties and avoid touching the plants when they are wet to prevent the spread of disease. Aphids and whiteflies can also cause problems.

Peppers

Capsicum

Features: bushy annual; attractive foliage; white flowers; colorful fruit
Height: 6–24" **Spread:** 12–18"

The variety of sweet and chili peppers is nothing short of remarkable. Among the many different shapes, colors and flavors of peppers, there is sure to be one or two that you'll fall in love with. A cool summer can really put a damper on fruiting, which can make hot pepper growing, in particular, a bit tricky in gardens west of the Cascades and very difficult near the coast.

Starting

Peppers need warmth to germinate and grow, and they take a while to mature, so it is best to start them indoors 6 to 10 weeks before the last frost date. If you don't have ideal light conditions, your seedlings may become stretched out and do poorly even when moved to the garden. In this case, you may prefer to purchase started plants; many varieties are available at garden centers and nurseries.

Growing

Peppers grow best in **full sun** in a warm location. The soil should be **average to fertile, moist** and **well drained**. Mulch to ensure the soil stays moist because peppers need to be planted in a fairly hot garden location. Most gardens will be hot enough to grow sweet peppers, but chili peppers need hotter weather to bear fruit. Try a dark mulch in a hot spot, or grow the plants in clay pots on a sunny patio against a south- or west-facing wall.

Harvesting

Peppers can be picked as needed once they are ripe. Chili peppers can also be dried for use in winter.

Tips

Pepper plants are neat and bushy. Once the peppers set and begin to ripen, the plants can be very colorful as the bright red, orange and yellow fruit contrasts beautifully with the dark green foliage. All peppers are good additions to container plantings. Chili peppers in particular are useful in containers because they usually have the most interesting fruit shapes, and the containers can be moved indoors or to a sheltered spot to extend the season if needed. Some of the smaller chili peppers also make interesting houseplants for warm, sunny windows.

Capsaicin is the chemical that gives peppers their heat. Peppers may be rated on a scale of 1–10 or by Scoville units. Sweet peppers have about 100 units while habaneros have up to 350,000.

Recommended

C. annuum is the most common species of sweet and hot peppers. Plants are bushy with dark green foliage. Flowers are white, and peppers can be shades of green, red, orange, yellow, purple or brown. Cultivars of sweet peppers include **'Yankee Bell,' 'Golden Bell,' 'Red Mini Bell,' 'King of the North'** and **'Purple Beauty.' 'North Star'** is an early standard bell pepper resistant to mosaic virus. Cultivars of chili peppers include **'Anaheim,' 'Cayenne,' 'Jalapeno,' 'Scotch Bonnet'** and **'Thai,'** just to name a few.

***C. chinense* 'Habanero'** is one of the hottest chili peppers. It is native to the Caribbean.

Problems and Pests

Rare problems with aphids and whiteflies can occur.

C. annuum (above)

Poppies

Papaver

Features: red, pink, white or purple, single or double flowers; edible seeds
Height: 1–4' **Spread:** 8–18"

Poppy seeds are a frequent addition to baked goods. Remarkably easy to grow, poppies also add abundant color to your garden.

P. somniferum (photos this page)

Starting

Direct sow poppies in spring. Several successive smaller sowings will give you a longer flower display, but it's not necessary if you are growing poppies for the seeds. The seeds are very small; mix them with fine sand before you plant them. Because they need light to germinate, don't cover the seeds when you plant them.

Growing

Poppies grow best in **full sun**. The soil should be **fertile, sandy, humus rich** and **well drained**. Good drainage is essential.

Harvesting

The seeds are ready to be harvested when the pods begin to dry and the seeds rattle in the pods when you shake them gently. Cut the heads off and shake the seeds into a paper bag. Let them dry in the paper bag and then store them in a container.

Tips

Poppies work well in mixed borders where other plants are slow to fill in. Poppies will fill empty spaces early in the season then die back over summer, leaving room for other plants. They self-seed freely and are likely to continue to spring up in your garden year after year.

Recommended

P. somniferum (opium poppy) forms a basal rosette of foliage above which leafy stems bear red, pink, white or purple flowers. Large, blue-green seedpods follow the flowers. Propagation of the species is restricted in many countries because of its narcotic properties, but several acceptable cultivars are available for ornamental and culinary purposes. Blue-, white- or brown-seeded varieties are available.

Problems and Pests

Poppies rarely suffer from any problems.

Growing this plant here is unlikely to give you any trouble with law enforcement. Poppies need intense, prolonged heat to develop enough of the compounds necessary to produce extractable narcotics, and the summer weather is not hot enough in Washington or Oregon to grow dangerous poppies.

Potatoes

Solanum

Features: bushy annual; leafy, rounded habit; pink, purple or white flowers
Height: 18–24" **Spread:** 12–24"

Potatoes were cultivated in South America for centuries and were introduced to Europe by the Spanish. They were only introduced to North America after European immigrants brought them here.

Starting

Sets of seed potatoes (small tubers) can be purchased and planted in spring a few weeks before the last frost date as long as the soil isn't cold and wet. Young plants can tolerate a light frost, but not a hard freeze. The seed potatoes can be cut into smaller pieces as long as each has an "eye," the dimpled spot from which the plant and roots grow. Each piece needs 12–18" of space to grow.

Growing

Potatoes prefer **full sun** but tolerate some shade. The soil should be **fertile, humus rich, acidic, moist** and **well drained**, though potatoes adapt to most growing conditions and tolerate both hot and cold weather. Mound soil up around the plants to keep the tubers out of the light as they develop. Potatoes need low nitrogen but high phosphorus. Mixing bone meal into the soil will increase your harvest, especially west of the Cascades where the soil

All parts of the potato plant are poisonous except the tubers, and they can become poisonous if they are exposed to light. Green flesh is a good indication that your potatoes have been exposed to light. To protect your potatoes, mound soil around the plants, 1" or so per week, from mid-summer to fall. A straw mulch also effectively shades the developing tubers.

is naturally acidic. Avoid using high-nitrogen manure on your potatoes.

Harvesting

The tubers begin to form around the same time the plants begin to flower, usually sometime in August. You can dig up a few tubers at a time from this point on as you need them. The remaining crop should be dug up in fall once the plants have withered, but before the first hard frost. Let them dry for a few hours on the soil, then brush the dirt off and store the tubers in a cold, dark place. You can even save a few of the smaller tubers for planting the following spring.

Tips

These large, bushy plants with white, pink or light purple flowers are good fillers for an immature border and are excellent at breaking up the soil in newer gardens.

Recommended

S. tuberosum is a bushy, mound-forming plant. It bears tiny, exotic-looking, white, pink or light purple flowers in late summer. There are many varieties of potatoes available. They can have rough or smooth, white, yellow, brown, red or blue skin and white, yellow, purple or blue flesh. A few popular varieties include **'All-Blue,'** with smooth, blue skin and light purple-blue flesh; **'Superior,'** with smooth, brown skin

and white flesh; **'Red Gold,'** with smooth, reddish orange skin and yellow flesh; and **'Yukon Gold,'** with smooth, light beige skin and yellow flesh. **'Butte'** is a good russet potato for baking and winter storage.

Problems and Pests

Potatoes are susceptible to a variety of diseases, including scab. Avoid planting them in the same spot two years in a row. Potato beetle is the most troublesome insect pest.

Purchase seed sets of a variety that interests you rather than trying to grow potatoes bought from the grocery store. They may have been treated to prevent sprouting, or they may be poorly suited to grow where you live.

Radishes

Daikon Radish

Raphanus

Features: annual; rosette of foliage; fast growing **Height:** 6–8" **Spread:** 4–6"

Radishes are grouped into two categories. Spring or salad radishes, which include the familiar round, red radishes and the cylindrical radishes, are used in salads. Winter radishes, which include Oriental or daikon radishes and Spanish radishes, are eaten raw or cooked and can be pickled. The spicy pods of some Oriental radishes are also popular in salads.

Starting

Direct sow seed in spring as soon as the soil warms up a bit and can be worked. Plants tolerate light frost. Successive, smaller plantings can be made every couple of weeks to ensure a steady supply of radishes. Daikon and other radishes that can be stored for winter can also be started in mid-summer to be harvested in fall.

Growing

Radishes grow well in **full sun** or **light shade**. The soil should be of **average fertility, loose, humus rich, moist** and **well drained**. Heavy or rocky soils cause the roots to be rough, woody and unpleasant tasting.

Harvesting

Spring radishes should be picked and eaten as soon as the roots develop. The flavor and texture deteriorates quickly if they are left in the ground or stored for too long.

Daikon and Spanish radishes are usually started in summer to be ready for harvest in late fall. They can be stored, like carrots, in moist sand in a cool, dry location. They can also be pickled.

Radishes tend to bolt in hot weather, and the roots develop an unpleasantly hot flavor. Choose icicle types for your summer growing because they are more tolerant of hot weather than the round, red varieties.

Tips

Easy to grow and a great addition to salads, radishes are also great nurse plants. If planted with mature plants or those that are small-seeded or slow to germinate, such as parsnips and carrots, radishes will shade out weeds and reduce evaporation. Radishes sprout and mature quickly; some varieties are ready to harvest within a month.

Because of their leafy, low-growing habit, these plants make interesting edging plants for borders and are unique additions to container plantings. Use radishes for a quick-growing spring display that will be replaced by other plants once the weather warms up.

Recommended

R. sativus forms a low clump of leaves. The edible roots can be long and slender or short and round; the skin can be rosy red, white or black. The cool-season radish **'Cherry Belle'** can be planted as soon as the soil can be worked in spring, and you can harvest it in as little as 22 days.

Problems and Pests

Flea beetles and cabbage maggots are common problems for radishes. Cover the crops with a row cover like Reemay to keep out these pests.

Radishes are related to cabbage, broccoli and mustard. They were grown by the ancient Greeks and Romans.

Raspberries & Blackberries

Rubus

Features: thicket-forming shrub; long, arching canes; spring flowers; summer fruit
Height: 3–10' **Spread:** 4–5' or more

The sweet, juicy berries of these shrubs are popular for use in pies, jams and other fruity desserts, but they are just as delicious when eaten fresh. Thickets are often pruned and staked to form neat rows but can be left to spread freely if you have the space.

Starting

Bare-root canes should be purchased in late winter or early spring and should be planted while they are still dormant. Container-grown plants are often available all season, but they may not establish as well.

Growing

Raspberries and blackberries grow well in **full sun, light shade** or **partial shade**, though the best fruiting occurs in full sun. The soil should be of **average fertility, humus rich, moist** and **well drained**. These plants prefer a location **sheltered** from strong winds. Blackberries, in particular, are prone to winter damage, but blackberries are more adapted to a wide range of soil types and will even tolerate some poor drainage. Raspberries should be grown in raised beds if drainage is not perfect, and they respond well to a mulch of well-rotted manure in early spring.

Prune out some of the older canes each year once plants become established to keep plants vigorous and to control their spread.

Raspberries come in July-bearing types that should be pruned in fall by removing the canes that have already fruited. There are also 'Everbearing' raspberries, which can be mowed to the ground each winter to get a fresh

Spread excess fruit out on a paper-lined cookie sheet and freeze it. Transfer the fruit to a plastic bag once it is frozen.

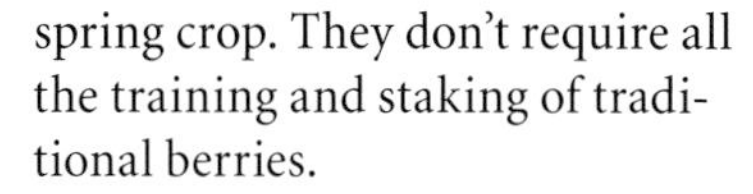

spring crop. They don't require all the training and staking of traditional berries.

Blackberries have very long, flexible canes. They can stand freely without staking, but they may take up less room if they are loosely tied to a supportive structure such as a stout post or fence.

Harvesting

Pick fruit as soon as it is ripe in mid- to late summer. All the fruit does not ripen at once, and you can harvest it for a month or more. Some raspberry varieties are ever-bearing and produce fruit in flushes from mid-summer through fall.

Tips

These shrubs form rather formidable thickets and can be used in shrub and mixed borders, along fences and as hedges.

Recommended

R. fruticosus (blackberries, brambles) forms a thicket of thorny stems or canes. Canes can grow up to 10' long, and thickets can spread 5' or more. White, or occasionally pink, late-spring or early-summer flowers are followed by red or black berries in late summer. Thornless varieties such as **'Apache'** and **'Arapaho'** are available and grow without the need of a trellis, while **'Triple Crown'** and **'Chester'** are thornless blackberries with a more vining habit that will need support (Zones 5–8). **'Tayberries'** are from Scotland and do especially well in the cool summer zones close to the coast.

R. idaeus (raspberries) forms a thicket of bristly stems or canes. Canes grow 3–5' long, and thickets can spread to 4' or more. White, spring flowers are followed by red, yellow, black or purple fruit in mid-summer. Raspberries fall into two categories: summer-bearing and ever-bearing. Although raspberry shrubs are perennial, the canes are biennial, generally growing the first year and producing fruit the second. In the third season, the canes die back. Ever-bearing canes begin to fruit late in the season of their first fall, then again the second summer. They are suitable west of the Cascades where fall is long and warm. Summer-bearing varieties are more suitable for shorter season areas. (Zones 3–8)

Problems and Pests

Problems with anthracnose, powdery mildew, rust, fire blight, leafhoppers and caterpillars can occur. Berry bushes wear out after 10 to 15 years, so replace them with fresh plants.

Rhubarb

Rheum

Features: clump-forming perennial; red or green stems; large, deeply veined leaves; spikes of flowers in summer **Height:** 2–4' **Spread:** 3–6'

Like tomatoes, rhubarb treads the fine line between fruit and vegetable. We usually eat it in fruit dishes and think of it more as a fruit, but rhubarb is actually a vegetable.

Starting

Rhubarb can be started from seed sown directly in the garden, or crowns can be purchased in spring. If you've planted crowns, you can begin harvesting the second summer. Seeded plants may take three or more years before they produce a harvest. If friends or neighbors have a rhubarb plant, ask them if they will give you a division from their plant.

Growing

Rhubarb grows best in **full sun**. The soil should be **fertile, humus rich, moist** and **well drained**, but this plant adapts to most conditions. Gently work some compost into the soil around the rhubarb each year, and add a layer of compost mulch. Fertile soil encourages more and bigger stems. Rhubarb can stay in one spot for many years but will remain more vigorous and productive if divided every eight or so years. Dig the soil over and work some more compost in when you divide the plant. Roots can be broken into smaller sections; each piece will sprout as long as it has an "eye."

Only the stems of rhubarb are edible. The leaves contain oxalic acid in toxic quantities.

Inedible ornamental rhubarb (below)

Harvesting

Harvest the stems by pulling them firmly and cleanly from the base of the plant. The leaves can then be cut from the stems with a sharp knife and composted or spread around the base of the plant to conserve moisture, suppress weed growth and return nutrients to the soil. Don't remove more than half the stems from the plant in one year. Rhubarb's flavor is better earlier in summer, and harvesting generally stops by early July when the stems start to become dry, pithy and bitter.

Tips

Sadly, this stunning plant is often relegated to back corners and waste

Although the flowers are quite interesting and attractive, they are often removed to prolong the stem harvest.

areas in the garden. With its dramatic leaves, bright red stems and intriguing flowers, rhubarb deserves a far more central location. A single rhubarb plant makes a dramatic centerpiece in the middle of a half whiskey barrel. Add trailing and upright flowers around the rhubarb.

Recommended

R. rhabarbarum and ***R.* x *hybridum*** form large clumps of glossy, deeply veined, green, bronzy or reddish leaves. The edible stems can be green, red or a bit of both. Spikes of densely clustered red, yellow or green flowers are produced in mid-summer. Popular varieties include **'Colossal,'** with huge leaves and stems, **'Valentine'** and **'Victoria.'** (Zones 2–8)

Problems and Pests

Rhubarb rarely suffers from any problems.

Rosemary

Rosemarinus

Features: fragrant, evergreen foliage; bright blue, sometimes pink flowers
Height: 8"–4' **Spread:** 1–4'

The needle-like leaves of this fragrant little shrub are used to flavor a wide variety of foods, including chicken, pork, lamb, rice, tomato and egg dishes.

Starting

Specific varieties of this plant can be purchased from garden centers, nurseries and specialty growers. Seed for the species is available but usually not for specific varieties. It can be started indoors in late winter.

Growing

Rosemary prefers **full sun** but tolerates partial shade. The soil should be of **poor to average fertility** and **well drained.** West of the Cascades, rosemary will survive winter outdoors in a protected spot. Grow it in a rock garden or raised bed so it will not drown in the wet winter weather.

Harvesting

Leaves can be picked as needed for use in cooking.

Tips

Harvest stems of rosemary to use as spears for meat and vegetable kebabs on the barbecue. You can even use the woody stems from rosemary plants that have been pruned back in early summer.

Recommended

R. officinalis is a dense, bushy, evergreen shrub with narrow, dark green leaves. The habit varies somewhat between cultivars, from strongly upright to prostrate and spreading. Flowers are usually shades of blue, but pink-flowered cultivars are available. Cultivars that are hardy in a sheltered spot in zone 6 with winter protection are also available. Plants rarely reach their mature size when grown in containers. (Zone 8)

Problems and Pests

Aphids and whiteflies can be a problem on plants overwintering indoors.

To overwinter a container-grown plant, keep it in very light or partial shade outdoors in summer, then put it in a sunny window indoors for winter and keep it well watered, allowing it to dry out slightly between watering.

Rutabagas

Brassica

Features: rosette-forming biennial; edible roots; blue-green foliage **Height:** 12–24" **Spread:** 4–8"

One of the tastiest of the root vegetables, rutabagas are not as popular as they deserve to be. Try them in soups and stews, to which they add a warm, buttery, sweet flavor. They are also delicious when roasted with other root vegetables.

Starting

Sow seeds directly into the garden in spring. Keep the seedbed moist until the plant germinates.

Growing

Rutabagas prefer to grow in **full sun**. The soil should be **fertile, moist** and **well drained**. The roots can develop discolored centers in boron-deficient soil. Work agricultural boron into the soil if needed. Rutabagas need an acidic soil so do best west of the Cascades.

Harvesting

Rutabagas are harvested in fall and stored for winter use. They can be used as soon as the roots are plump and round, but they can also be left in the ground so the first few fall frosts sweeten the roots. Dig them up, cut the greens off and let the roots dry just enough so the dirt can be brushed off, then store them in moist sand in a cold, frost-free location for winter.

Tips

Rutabagas grow large, bushy clumps of blue-green foliage and can be included in the middle of a border where, though they aren't particularly showy, they provide an attractive, contrasting background for other plants.

Recommended

B. napus (rutabaga, swede, winter turnip) forms a large clump of blue-green leaves. The leaves are not edible. The roots are most often white with purple tops and have yellow flesh. A few varieties are usually available. **'American Purple Top'** is a popular variety. **'Laurentian'** and its many variations are commonly available and are popular because they store well. Choose a variation that is clubroot resistant, such as **'Marian.'**

Problems and Pests

Cabbage root maggots, cabbage worms, aphids, flea beetles, rust, fungal diseases, downy mildew, powdery mildew and clubroot are possible problems.

The rutabaga originated in Europe and has been in cultivation for over 4000 years.

Sage

Salvia

Features: fragrant, decorative foliage; blue or purple, summer flowers
Height: 12– 24" **Spread:** 18–36"

Sage is perhaps best known as a flavoring for stuffing, but it has a great range of uses, including in soups, stews, sausages and dumplings.

Starting

The species can be started indoors in late winter or early spring, but you will have to purchase specific varieties from a garden center.

Growing

Sage prefers **full sun** but tolerates light shade. The soil should be of **average fertility** and **well drained**. This plant benefits from a light mulch of compost each year. It tolerates drought once established.

Harvesting

Pick leaves as you need them for fresh use. They can be dried or frozen in late summer and fall for winter use. A winter mulch of straw often provides enough protection for sage to survive up to a zone 3 winter.

Tips

Sage is a good plant for a border, adding volume to the middle or as an attractive edging or feature plant near the front. Sage can also be grown in mixed planters and is often sold as a foliage plant for fall container gardens.

Recommended

S. officinalis is a woody, mound-forming perennial with soft, gray-green leaves. Spikes of light purple or blue flowers appear in early and mid-summer. Many cultivars with attractive foliage are available, including the yellow-leaved **'Golden Sage,'** the yellow-margined **'Icterina,'** the purple-leaved **'Purpurea'** and the purple, green and cream variegated **'Tricolor,'** which also has a pink flush to the new growth.

Problems and Pests

Sage rarely suffers from any problems, but it can rot in wet soil.

Sorrel

Rumex

Features: low-growing perennial; tangy, edible leaves **Height:** 6–36" **Spread:** 12–24"

Sorrel can be classified somewhere between a salad green and a herb. The tangy, sour leaves are a perfect addition to a salad of mixed greens, but the leaves can also be used to flavor soups, marinades and egg dishes.

Oxalic acid gives sorrel leaves their flavor, and though they are safe to eat, in large quantities they can cause stomach upset.

R. acetosa

Starting

Sow the seed directly in the garden in fall or spring. You can start the seed early indoors, but it isn't really necessary because the plant grows quickly.

Growing

Sorrel grows well in **full sun, light shade** or **partial shade**, especially in hotter parts of Washington and Oregon. The soil should be **average to fertile, acidic, humus rich, moist** and fairly **well drained**, but this plant is adaptable. Mulch to conserve moisture.

Divide sorrel every three or four years to keep the plant vigorous and the leaves tender and tasty. This plant self-seeds if flowering spikes are not removed.

Harvesting

Pick leaves as needed in spring and early summer. Remove flower spikes as they emerge to prolong the leaf harvest. Once the weather warms up and the plant goes to flower, the leaves lose their pleasant flavor. If you cut the plant back a bit at this point, you will have fresh leaves to harvest in late summer and fall when the weather cools again.

Tips

A tasty and decorative addition to the vegetable, herb or ornamental garden, French sorrel also makes an attractive groundcover and can be included in mixed container plantings.

Recommended

R. acetosa (garden sorrel, broad leaf sorrel) is a vigorous, clump-forming perennial. The inconspicuous flowers are borne on a tall stem that emerges from the center of the clump. This plant grows 18–36" tall and spreads about 12". (Zones 4–8)

R. scutatus (French sorrel, buckler leaf sorrel) forms a low, slow-spreading clump of foliage. The leaves are stronger tasting but are not as plentiful as those of garden sorrel. This selection keeps its flavor better in warm weather than garden sorrel does. It grows 6–18" tall and spreads up to 24". (Zones 4–8)

Problems and Pests

Sorrel is generally problem free. Snails, slugs, rust and leaf spot can cause problems occasionally.

Spinach

Spinacia

Features: clump-forming annual; smooth or crinkled, edible leaves
Height: 12–18" **Spread:** 6–10"

Nutritious and versatile, spinach is useful in a wide variety of dishes. And it's effortless to grow, so you might want to consider adding this popular leafy plant to your garden, even if you don't grow any other vegetables.

Starting

Direct sow spinach in spring as soon as the soil can be worked. Young plants can tolerate a light frost, but if the temperature is expected to fall below 23° F, you should cover them. Several successive sowings in spring and again in mid- to late summer will provide you with a steady supply of tender leaves.

Growing

Spinach grows well in **full sun** or **light shade** and prefers cool weather and a cool location. The soil should be **fertile, moist** and **well drained**. Add a layer of mulch to help keep soil cool because this plant bolts in hot weather. Many gardeners west of the Cascades won't have to worry about bolting because our cool summer nights are ideal for growing spinach. Bolt-resistant varieties are also available.

Harvesting

Pick leaves, as needed, a few at a time from each plant. The flavor tends to deteriorate as the weather heats up and the plant matures and goes to flower.

The unrelated New Zealand spinach (Tetragonia expansa) *is an excellent and interesting alternative to regular spinach. The leaves can be used in the same way, but the plant is far more heat resistant, grows upright and branching in habit and can be planted in summer, as regular spinach begins to fade. It will grow quickly in the heat of summer and be ready for a late-summer and fall harvest.*

Tips

Spinach's dark green foliage is attractive when mass planted and provides a good contrast for brightly colored flowers. Try it in a mixed container that you keep close to an entryway to make harvesting more convenient.

Recommended

S. oleracea forms a dense, bushy clump of glossy, dark green, smooth or crinkled (savoyed) foliage. Plants are ready for harvest in about 45 days. **'Bloomsdale'**

produces dark green, deeply savoyed foliage. Many strains of this cultivar are also available. **'Bordeau'** is a smooth-leaved variety with dark red veins. **'Olympia'** is a deep green variety you can harvest year-round and is mildew resistant. **'Tyee'** is a semi-savoy variety that is very bolt resistant.

Problems and Pests

Avoid powdery and downy mildew by keeping a bit of space between each plant to allow for good air flow.

Squash

Cucurbita

Features: trailing or mounding annual; large lobed, decorative leaves; colorful flowers and fruit **Height:** 18–24" **Spread:** 2–10'

Squash are generally grouped as summer squash and winter squash, reflecting when we eat the squash more than any real difference in the plants themselves. All squash develop hardened rinds in fall if left to mature. The squash that keep the longest and have the best flesh taste and texture when mature are winter squash. Summer squash are tender and tasty when they are immature but tend to become stringy and sometimes bitter when they mature, and they don't keep as well.

Starting

Start seeds in peat pots indoors six to eight weeks before the last frost date. Keep them in as bright a location as possible to reduce stretching. Plant out or direct sow after the last frost date and once the soil has warmed up.

Growing

Squash grow best in **full sun** but tolerate light shade from companion plants. The soil should be **fertile, humus rich, moist** and **well drained**. Mulch well to keep the soil moist. Put mulch or straw under developing fruit of pumpkins and other heavy winter squash to protect the skin while it is tender. Squash generally need a long, warm summer to develop well. Gardeners in cooler areas will want to choose species and cultivars that mature in a shorter season.

Harvesting

Summer squash are tastiest when picked and eaten young. The more you pick, the more the plants will produce. Cut the fruit cleanly from the plant, and avoid damaging the leaves and stems to prevent disease and insect problems.

Winter squash should be harvested carefully, to avoid damaging the skins, just before the first hard frost. Allow them to cure in a warm dry place for a few weeks until the skins become thick and hard. They can then be stored in a cool dry place, where they should keep all winter. Check them regularly to be sure they aren't spoiling.

Tips

Mound-forming squash, with their tropical-looking leaves, can be added to borders as feature plants.

C. pepo (above & below)

Small-fruited trailing selections can be grown up trellises. The heavy-fruited trailing types will wind happily through a border of taller plants or shrubs. All squash can be grown in containers, but the mound-forming and shorter-trailing selections are usually most attractive; the long-trailing types end up as a stem that leads over the edge of the container.

Recommended

Squash plants are generally similar in appearance, with medium to large leaves held on long stems. Plants are trailing in habit, but some form only very short vines, so they appear to be more mound forming. Bright yellow, trumpet-shaped male and female flowers are borne separately but on

the same plant. Female flowers can be distinguished by their short stem and by the juvenile fruit at the base of the flower. Male flowers have longer stems.

There are four species of squash commonly grown in gardens. They vary incredibly in their appearance and can be smooth or warty, round, elongated or irregular. In color, they can be dark green, tan, creamy white or bright orange, and solid, stripy or spotted. The size ranges from tiny, round zucchini that would fit in the palm of a child's hand to immense pumpkins that could hold two or three children. Experiment with different types to find which ones grow best in your garden and which ones you like best.

C. maxima includes buttercup squash and hubbard. These plants generally need 90 to 110 days to mature. They keep very well, often longer than any other squash, and have sweet, fine-textured flesh. **'Blue Ballet'** will mature in 90 days so is good for short summer areas.

C. mixta includes cushaw squash and is not as well known. These plants generally take 100 days or more to mature. Some are grown for their edible seeds, while others are used in baking and are good for muffins, loaves and pies.

C. moschata includes butternut squash and generally needs 95 or more days to mature. These squash keep well and are popular baked or for soups and stews. **'Early Butternut'** is best for cool summer gardens.

C. pepo is the largest group of squash and includes summer squash, such as zucchini, and many winter squash, such as pumpkins, acorn squash, spaghetti squash, dumpling squash and gourds. Summer squash are ready to harvest in 45 to 50 days, and the winter squash in this group take from 70 to 75 days for acorn and spaghetti squash to 95 to 120 days for some of the larger pumpkins. **'Sungreen'** is a virus-resistant squash, and **'Gold Rush'** has compact, beautiful fruit.

Problems and Pests

Problems with mildew, cucumber beetles, stem borers, bacterial wilt and whiteflies can occur. Ants may snack on damaged plants and fruit, and mice will eat and burrow into squash for the seeds in fall.

Don't worry if some of your summer squash are too mature or your winter squash are not mature enough. Summer squash can be cured and will keep for a couple of months. They are still useful for muffins and loaves. Immature winter squash can be harvested and used right away; try them stuffed, baked or barbecued.

C. pepo (photos this page), female flower (above)

Stevia

Stevia

Features: green, hairy annual or perennial; toothed leaves; white, tubular flowers
Height: 24–36"
Spread: 12–24"

This remarkable plant contains steviol glycosides, compounds that are up to 300 times sweeter than sugar but without the calories. Stevia powder is available at health food stores, but you can easily grow your own to help reduce your processed sugar intake.

Starting

Stevia has poor germination rates, so it is easier to start with a new plant. Stevia is not frost tolerant so set out after all danger of frost is passed.

Growing

Stevia prefers **full sun** or **partial shade**. The soil should be **moist, fertile** and **well drained**, especially in western Washington where heavy rain can rot the roots. The plant has shallow roots, so a light mulch will help protect them. Feed with a low nitrogen plant food.

Pinch out emerging flowers to allow more time for glycoside accumulation.

Harvesting

Harvest the leaves before the plant flowers to ensure the highest concentration of glycosides. Fresh leaves can be used immediately, but the sweetness increases exponentially when dried. Cut the top 4" off the plants and then remove the leaves. Dry them on a screen and then grind or chop the dry leaves. Store in an airtight container for use as sweetening agent in tea, coffee and baked goods, or make a slurry with chopped leaves and water in a blender. Once dry, store in an airtight container for use in tea and coffee and baking.

Tips

Integrate stevia into mixed beds and borders as edging, and plant it in group for more impact. If you plant it in a container, use a lightweight potting soil and bring it indoors for winter. One to three plants should supply the average person with enough leaves all year. To propagate, take cuttings in fall after you harvest the leaves. A stem cutting can be rooted with the aid of a rooting hormone.

Recommended

S. rebaudiana can grow 24–36" tall and 12–24" wide. Hairy stems are covered in dark green, toothed leaves. White, tubular flowers are produced in mid-summer. The leaves are not aromatic but are sweet to the taste; dried leaves are even sweeter.

Problems and Pests

Stevia is not afflicted with pests or problems.

Strawberries

Fragaria

Features: spreading perennial; soft, bright green leaves; white, sometimes pink flowers; bright red, edible fruit **Height:** 6–12" **Spread:** 12" or more

Many of these plants, with their pretty little white flowers, spread vigorously by runners. Long shoots spread out from the parent plant, and small baby plants grow at the tips. Purchasing just a few plants will provide you with plenty of fruit-producing plants by the end of summer.

F. vesca

Starting

Some selections can be started indoors about 12 weeks before you plan to plant them outside. Other selections are only available as crowns or plants. Plant them outdoors around the last frost date. They tolerate light frosts.

Growing

Strawberries grow well in **full sun** or **light shade**. The soil should be **fertile, neutral to acidic, moist** and **well drained**.

Harvesting

Pick strawberries as soon as they are ripe. Some types produce a single large crop of fruit in early summer, and others produce a smaller crop throughout most or all of summer.

Tips

Strawberries make interesting, tasty and quick-growing groundcovers. They do well in containers, window boxes and hanging baskets. The selections that don't produce runners are also good for edging beds.

Recommended

F. chiloensis (Chilean strawberry), ***F. vesca*** (wild strawberry, alpine strawberry) and ***F. virginiana*** (Virginia strawberry) have been crossed to form many hybrids. Similar in appearance, they generally form a low clump of three-part leaves and may or may not produce runners. Flowers in spring are followed by early- to mid-summer fruit. Some plants produce a second crop in fall, and others produce fruit all summer. Fruit of wild or alpine strawberries is smaller than the fruit of the other two species. Popular cultivars include **'Seascape'** and **'Puget Summer'** for cool summer areas near the coast, and the disease-resistant **'Earliglo.' 'Tristar'** and **'Eversweet'** are everbearing or day neutral varieties. (Zones 3–8)

Problems and Pests

The fruit is susceptible to fungal diseases, so mulch to protect it. Some leaf spot, spider mite problems and wilt can occur. When planting, do not set the crowns too deep to avoid root rot problems in wet weather.

Sunflowers & Sunchokes

Helianthus

Features: daisy-like, yellow, orange, red, brown, cream or bicolored flowers, typically with brown, purple or rusty red centers; edible seeds; edible tubers **Height:** 6–15' **Spread:** 1–4'

Sunflowers make wonderful companions for other seed-bearing plants. Sunchokes are a native species that develop edible tubers that taste like artichokes, radishes or waterchestnuts, depending on who is describing the flavor.

Starting

Sunflowers and sunchokes can be sown directly into the garden in spring around the last frost date. Water well until the plants become established. Sunchokes are perennial and will grow back each year as long as you leave some of the tubers in place in fall.

Growing

Sunflowers and sunchokes grow best in **full sun**. The soil should be of **average fertility, humus rich, moist** and **well drained**, though plants adapt to a variety of conditions. They become quite drought tolerant as summer progresses.

Harvesting

Sunflower seeds are ready to harvest when the flower has withered and the seeds are plump. You may need to cover the flower heads with a paper bag or net to prevent birds from eating all the seeds before you get to them.

Sunchoke tubers are usually ready to harvest around the time of the first frost in fall. They should be stored in a cool, dry, well-ventilated area.

Tips

Sunflowers make a striking addition to the back of a border and along fences and walls. These tall plants may need staking in windy or exposed locations.

Sunchokes can be boiled, baked, fried, steamed, stewed or eaten raw. They cook more quickly than potatoes do and become mushy if overcooked.

Recommended

H. annuum (sunflower) can develop a single stem or many branches. The large-flowered sunflowers usually develop a single stem. Many of the ornamental sunflowers offered in gardening catalogs have edible seeds. **'Russian Mammoth,'** a tall cultivar with large, yellow flowers, is one of the most popular for seed production.

Sunflowers are a great way to invite birds into your garden if you don't mind not having the seeds for yourself. The birds will feast on garden pests as well as the seeds the flowers provide.

H. annuum (photos this page)

H. tuberosus (sunchoke, Jerusalem artichoke) is a tall, bushy, tuberous perennial. It grows 6–10' tall and spreads 2–4'. Bright yellow flowers are produced in late summer and fall. (Zones 2–8)

Problems and Pests

Plants are generally problem free, but keeping birds away from the seeds until they are ready to be picked can be troublesome.

Sunchokes store energy in the tubers as a carbohydrate called inulin, rather than starch. They are filling, but the energy is not readily absorbed by the body and does not affect blood sugar levels, making these tubers useful for diabetics and dieters.

Tarragon

Artemisia

Features: narrow, fragrant leaves; airy flowers **Height:** 18–36" **Spread:** 12–18"

The distinctive licorice-like flavor of tarragon lends itself to a wide variety of meat and vegetable dishes and is the key flavoring in Bernaise sauce.

Starting

The tastiest tarragon is French tarragon, and it can only be propagated vegetatively. Seeds do not come true to type. Plants can be purchased at nurseries and garden centers and from specialty growers.

Growing

Tarragon grows best in **full sun**. The soil should be **average to fertile, moist** and **well drained**. Divide the plant every few years to keep it vigorous and for the best leaf flavor.

Harvesting

Pick leaves as needed to use fresh, and dry or freeze some in late summer for winter use.

Tips

Tarragon is not exceptionally decorative, but it does provide a good vertical presence. It can be included in a herb garden or mixed border, where the surrounding plants will support its tall stems.

Recommended

A. dracunculus* var. *sativa (French tarragon) is a bushy plant with tall stems and narrow leaves. Airy clusters of insignificant flowers are produced in late summer. (Zones 3–8)

Problems and Pests

Tarragon rarely suffers from any problems.

*Before purchasing a plant, chew a leaf to see if it has the distinct flavor it should have. French tarragon is preferred, whereas Russian tarragon (*A. d. *var.* dracunculoides*) is a more vigorous plant but has little of the desired flavor.*

Thyme

Thymus

Features: bushy habit; fragrant, decorative foliage; purple, pink or white, summer flowers **Height:** 8–16" **Spread:** 8–24"

Thyme is a popular culinary herb used in soups, stews, casseroles and roasts.

Thyme is a bee magnet when it is blooming. Pleasantly herbal thyme honey goes very well with biscuits.

Starting

Common thyme can be started indoors from seed four to six weeks before you plan to plant it outdoors. This plant also can be purchased at nurseries and garden centers and from specialty growers.

Growing

Thyme prefers **full sun**. The soil should be **neutral to alkaline**, of **poor to average fertility** and very **well drained**. Good drainage is essential. It is beneficial to work leaf mold and sharp limestone gravel into the soil to improve structure and drainage.

Harvesting

Pick leaves as needed or dry.

Tips

Thyme is useful for sunny, dry locations at the front of borders, between or beside paving stones, on rock walls and in containers. Once the plant has finished flowering, shear it back by about half to encourage new growth and to prevent it from becoming too woody.

Recommended

T.* x *citriodorus (lemon thyme) forms a mound of lemon-scented, dark green foliage. The summer flowers are pale pink. Cultivars with silver- or gold-margined leaves are available.

T. vulgaris (common thyme) forms a bushy mound of fragrant, dark green leaves. The summer flowers may be purple, pink or white. Cultivars with variegated leaves are available.

Problems and Pests

Thyme roots can rot in poorly drained, wet soils.

Tomatoes

Lycopersicon

Features: annual vine; bushy upright or trailing habit; fragrant foliage; yellow flowers; red, pink, orange or yellow, edible fruit **Height:** 18"–5' **Spread:** 18–36"

The colors and flavors available in the world of tomatoes is truly astounding. If you've only been eating tomatoes you've purchased from the grocery store, you are in for a huge surprise once you taste a tomato grown in your own garden.

Starting

Tomatoes can be started indoors six to eight weeks before the last frost date or can be purchased in spring from nurseries and garden centers.

Growing

Tomatoes grow best in **full sun**. The soil should be **average to fertile, humus rich, moist** and **well drained**. Keep tomatoes evenly moist to encourage good fruit production. Except for the very small bush selections, tomatoes tend to be quite tall and are prone to flopping over unless stakes, wire hoops, tomato cages or other supports are used.

Harvesting

Pick the fruit as soon as it is ripe. Tomatoes pull easily from the vine with a gentle twist when they are ready for picking.

Tips

Tomatoes are bushy and have both attractive little flower clusters and vibrantly colored fruit. Many of the selections grow well in containers and are included in patio gardens and hanging baskets.

Recommended

L. lycopersicum are bushy or vine-forming annuals with pungent, bristly leaves and stems. Determinate plants grow a specific height and are generally short enough to not need staking. Indeterminate plants continue to grow all summer and usually need staking. Clusters of yellow flowers are followed by fruit from mid-summer through to the first frost. Fruit ripen to red, orange, pink, yellow or purplish black and come in many shapes and sizes. Beefsteak tomatoes produce the largest fruit, and currant tomatoes produce the smallest fruit. Browse through a seed catalog to see the many offerings available. Try a few each year until you find your favorites. West of the Cascades,

When tomatoes plants were first introduced to Europe, they were grown as ornamental plants, not for their edible fruit.

'Early Girl' and **'Oregon Spring'** will ripen quickly in cool summer areas, and **'Legend'** is most resistant to blight. The small fruited cherry tomatoes are the easiest to grow and do well in containers. **'Sun Gold'** has small but bright gold fruit and an excellent flavor.

Problems and Pests

Problems with tobacco mosaic virus, aphids, fungal wilt, blossom end rot and nematodes can occur.

Tomatillos (Physalis ixocarpa) *are related to tomatoes, and the plants have similar cultural requirements. The fruit is encased in a delicate papery husk. It is ready to pick when the husk is loose and the fruit has turned from green to gold or light brown.*

Turnips

Brassica

Features: biennial grown as an annual; clump of edible, blue-green foliage; plump, edible root **Height:** 10–18" **Spread:** 6–8"

Tender with a delicate flavor, turnips should be eaten as soon as they mature for the best-tasting roots.

B. rapa (above & below)

Starting

Sow seeds directly into the garden in spring. Keep the seedbed moist until plants germinate. Several small, successive sowings will provide you with turnips for a longer time.

Growing

Turnips prefer **full sun**. The soil should be **fertile, acidic, moist** and **well drained**. The roots can develop discolored centers if the soil is boron deficient; agricultural boron can be worked into the soil if needed.

Harvesting

The leaves can be harvested a few at a time from each plant as needed and steamed or added to stir-fries. The roots should be harvested as soon as they are plump because they have the best flavor and texture when young. Turnips do not keep well when stored.

Tips

These plants form attractive clumps of foliage and can be added in small groups to borders, where the foliage provides a good background for flowering plants.

Recommended

B. rapa (turnip, summer turnip) is a biennial grown as an annual. It produces white or purple-shouldered, edible roots. The blue-green foliage is also edible. **'Hakurei'** and **'Purple Top White Globe'** are popular cultivars.

Problems and Pests

Cabbage root maggots, cabbage white butterfly larvae, aphids, rust, fungal diseases, downy mildew, powdery mildew and clubroot are possible problems.

Turnips, originally grown as animal fodder, only became popular for dinner in the 1600s.

Watermelons

Citrullus

Features: annual vine; climbing or trailing habit; yellow flowers; decorative dark green striped, pale green skinned fruit **Height:** 12" **Spread:** 5–10'

Tasty and juicy, watermelon is the ultimate summer treat. Make it an even bigger treat and try growing your own. It grows best in hot, humid weather, but plenty of short-season varieties are available for Washington and Oregon gardeners.

Starting

Watermelons require a long growing season and should be started indoors in individual peat pots four to six weeks before the last frost date. Started plants can also be purchased at garden centers and nurseries. Transplant them into the garden after the last frost date once the soil has warmed up.

Growing

Watermelons grow best in **full sun**. The soil should be **fertile, humus rich, moist** and **well drained**. They like plenty of water during the growth and early fruiting stages but, to intensify the flavor, should be allowed to dry out a bit once the fruit is ripening. Cover young plants with row covers to hold in warmth until July and use a heat-absorbing mulch on top of the soil as soon as you plant. A black plastic or silver mulch will trap and hold heat to encourage your watermelon plants to form fruit. In cool summer gardens near the coast, watermelons will not thrive.

Harvesting

A watermelon is generally ready to pick when the pale white area on the skin, where the fruit sits, turns yellow. Some experimentation may be required before you become adept at judging the ripeness of the fruit.

Tips

Small-fruited watermelons make attractive patio climbers, though the small fruit may need some support as it grows. These plants can be left to wind through ornamental beds and borders.

Recommended

C. lanatus takes 65 to 105 days to mature; the skin of the fruit may be light green with dark green stripes or solid dark green; the flesh may be red, pink, orange or yellow. Popular early maturing cultivars include **'Golden Crown,' 'Sugar Baby,' 'Yellow Doll'** and **'New Queen'** for short summer climates.

Problems and Pests

Problems with powdery mildew, *Fusarium* wilt, cucumber beetles and sap beetles can occur. Watermelon fruit blotch is a serious problem that can affect this plant.

Watermelon is native to tropical parts of Africa, but it was introduced to Asia and has been grown there for centuries.

Glossary

Acid soil: soil with a pH lower than 7.0

Alkaline soil: soil with a pH higher than 7.0

Annual: a plant that germinates, flowers, sets seeds and dies in one growing season

Basal leaves: leaves that form from the crown, at the base of the plant

Blanching: to deprive a plant or part of a plant of light, resulting in a pale color and usually a milder flavor

Bolting: when a plant produces flowers and seeds prematurely, usually rendering the plant inedible

Bract: a special, modified leaf at the base of a flower or inflorescence; bracts may be small or large, green or colored

Cross-pollination: the pollination of one plant by a closely related one; undesirable if the resulting seeds or fruit lack the expected qualities; beneficial if an improved variety results

Crown: the part of the plant at or just below soil level where the shoots join the roots

Cultivar: a cultivated plant variety with one or more distinct differences from the species, e.g., in flower color or disease resistance

Damping off: fungal disease causing seedlings to rot at soil level

Deadhead: removing spent flowers to maintain a neat appearance and encourage a long blooming season

Diatomaceous earth: an abrasive dust made from the fossilized remains of diatoms, a species of algae; the scratches it makes on insect bodies causes internal fluids to leak out, and the insects die of dehydration

Direct sow: to sow seeds directly into the garden

Dormancy: a period of plant inactivity, usually during winter or unfavorable conditions

Double flower: a flower with an unusually large number of petals

Drought resistant: can withstand drought for a long time

Drought tolerant: can withstand drought conditions, but only for a limited time

Genus: a category of biological classification between the species and family levels; the first word in a scientific name indicates the genus

Half-hardy: a plant capable of surviving the climatic conditions of a given region if protected from heavy frost or cold

Harden off: to gradually acclimatize plants that have been growing in a protected environment to a harsher environment

Hardy: capable of surviving unfavorable conditions, such as cold weather or frost, without protection

Humus: decomposed or decomposing organic material in the soil

Hybrid: a plant resulting from natural or human-induced cross-breeding between varieties, species or genera

Inflorescence: an arrangement of flowers on a single stem

Invasive: able to spread aggressively and outcompete other plants

Loam: a loose soil composed of clay, sand and organic matter, often highly fertile

Microclimate: an area of beneficial or detrimental growing conditions within a larger area

Mulch: a material (e.g., shredded bark, pine cones, leaves, straw) used to surround a plant to protect it from weeds, cold or heat and to promote moisture retention

Neutral soil: soil with a pH of 7.0

Node: the area on a stem from which a leaf or new shoot grows

Perennial: a plant that takes three or more years to complete its life cycle

pH: a measure of acidity or alkalinity; soil pH influences availability of nutrients for plants

Plantlet: a young or small plant

Potager: an ornamental kitchen garden, often laid out symmetrically with raised beds or low, hedge-edged beds

Rhizome: a root-like, food-storing stem that grows horizontally at or just below soil level, from which new shoots may emerge

Rosette: a low, flat cluster of leaves arranged like the petals of a rose

Runner: a modified stem that grows on the soil surface; roots and new shoots are produced at nodes along its length

Seed head: dried, inedible fruit that contains seeds

Self-seeding: reproducing by means of seeds without human assistance, so that new plants constantly replace those that die

Single flower: a flower with a single ring of typically four or five petals

Spathe: a leaf-like bract that encloses a flower cluster or spike

Species: the fundamental unit of biological classification; the entity from which cultivars and varieties are derived

Standard: a tree or shrub pruned to form a rounded head of branches at the top of a clearly visible stem

Subspecies (subsp.): a naturally occurring, often regional, form of a species, isolated from other subspecies but still potentially interfertile with them

Taproot: a root system consisting of one long main root with smaller roots or root hairs branching from it

Tender: incapable of surviving the climatic conditions of a given region and requiring protection from frost or cold

Tuber: the thick section of a rhizome bearing nodes and buds

Understory plant: a plant that prefers to grow beneath the canopies of trees in a woodland setting

Variegation: foliage that has more than one color, often patched, striped or bearing leaf margins of a different color

Variety (var.): a naturally occurring variant of a species

Index of Plant Names

Boldface type refers to the primary species accounts.

C

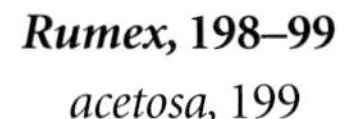

About the Authors

Marianne Binetti reaches over a million readers each week in the *Seattle Post-Intelligencer* with her syndicated garden column, which also is carried in 20 newspapers across Washington State. She contributes to several national magazines, including *Better Homes and Gardens, Woman's Day* and *Flower and Garden.* Marianne's light-hearted approach to gardening appeals to both television and radio audiences. She is a regular guest on HGTV and is also a popular speaker at garden centers and horticulture societies throughout Washington and Oregon.

Alison Beck has gardened since she was a child. She has a diploma in Horticulture and a degree in Creative Writing. Alison is the co-author of several best-selling gardening guides. Her books showcase her talent for practical advice and her passion for gardening.